Carbon Fibres in
Composite Materials

# Carbon Fibres in Composite Materials

R. M. GILL, B.Sc., Ph.D., C.Eng., A.R.I.C., A.M.Inst.F., F.I.Ceram.
*Formerly, Technical and Production Director,*
*Morganite Modmor Limited*

Distributed in the United States by
CRANE, RUSSAK & COMPANY, INC.
52 Vanderbilt Avenue
New York, New York 10017

THE BUTTERWORTH GROUP

ENGLAND
Butterworth & Co (Publishers) Ltd
London: 88 Kingsway, WC2B 6AB

AUSTRALIA
Butterworth & Co (Australia) Ltd
Sydney: 586 Pacific Highway Chatswood, NSW 2067
Melbourne: 343 Little Collins Street, 3000
Brisbane: 240 Queen Street, 4000

CANADA
Butterworth & Co (Canada) Ltd
Toronto: 14 Curity Avenue, 374

NEW ZEALAND
Butterworth & Co (New Zealand) Ltd
Wellington: 26–28 Waring Tayler Street, 1
Auckland: 35 High Street, 1

SOUTH AFRICA
Butterworth & Co (South Africa) (Pty) Ltd
Durban: 152–154 Gale Street


First published in 1972 by
Iliffe Books, an imprint of
the Butterworth Group

Published for The Plastics Institute
11 Hobart Place, London, SW1

ISBN 0 592 00069 9


Printed in England
by Page Bros (Norwich) Ltd, Norwich

# Preface

During the last 25 years there has been a rapid growth in the use of all types of resins, and in particular these materials have found an ever-widening use in structural applications. This period has also witnessed the establishment of resins containing reinforcing additives to enhance their mechanical properties. Recently, much progress has taken place in the development of fibres having exceptionally high mechanical strength and stiffness, which makes them eminently suitable for combining with resins to give reinforced materials of outstanding performance. Of these, carbon fibre is particularly noteworthy and is currently playing a leading role in the creation of a whole range of new materials of construction.

The technology surrounding carbon fibres and their associated composite materials is new and is advancing rapidly. The aim of this monograph is to present a broad review of the background and current achievements in this field. It must be recognised, however, that developments and improvements are continually taking place and much of the information presented here will undoubtedly have to be revised at an early date in the light of further experience.

The author would like to acknowledge the advice and assistance given by Dr J. Saunders, Mr J. Cunningham, and Mr I. B. Simpson. He would also like to thank Mr H. D. Blakelock and Mr J. C. Joiner for providing certain items of information. Additional data were provided by Mr L. P. Suffredini and Mr D. W. Gibson, for which the author is also grateful. Thanks are due to the Library staff and Technical Information Group of Morganite Research and Development Ltd for providing many of the references and to Miss M. A. Pullen for typing and help in checking the monograph.

London                                                                    R.M.G.

# Contents

# Part 1

# 1

# Fibres and composite materials

## 1.1  INTRODUCTION

Engineering achievement has always been closely associated with
the availability of suitable materials of construction. The twentieth
century has seen the development of highly sophisticated steels and
other metal alloys which have contributed enormously to the
advance of technology, affecting every walk of life. The tremendous
progress made in the performance of aircraft, for example, would
have been impossible without the use of the various new metals and
non-metallic materials.

Further progress in engineering will depend on the continued
development of all forms of constructional materials. For strong
lightweight structures, there is a very real need for materials having
greater specific stiffness and specific strength than are attainable
with existing metals. The specific strength of a constructional
material refers to the strength per unit weight; this is the all-
important criterion in connection with aircraft, since there is
nothing to be gained in producing a stronger structure if the weight
is increased in proportion. The same applies to the stiffness of a
structure.

Whilst the development of metal alloys is by no means complete,
it is becoming increasingly difficult to obtain significant improvements in their specific properties. Moreover, such developments are
almost invariably associated with greater sophistication during
manufacture, which is reflected in cost. The degree of improvement

sought by designers is much greater than appears to be forthcoming in metals in the foreseeable future unless there is some major advance in metallurgical technology.

In the last decade, much effort has been expended in the area of fibres and whiskers with a view to their possible application as materials of construction. For many years it has been known that certain elements and compounds in whisker form may exhibit outstanding strength and stiffness, owing to the fact that their structure contains fewer flaws or defects than are present in their bulk form. The figures given in *Table 1.1* show that many engineering metals have similar specific strengths, ranging from 20000 lbf/in$^2$ to 45000 lbf/in$^2$, whilst their specific stiffness values are remarkably constant at around $3 \cdot 5 \times 10^6$ lbf/in$^2$. In addition, although wood has a much lower stiffness than steel, it is also lighter and their specific stiffness values are therefore close to each other, as indicated in the table. Indeed, the specific strength and specific stiffness of almost all the bulk materials lie close together. However, it will be seen that the specific strength of glass fibre is almost ten times that of bulk metals, although the specific stiffness is only marginally better.

## 1.2   GLASS FIBRES AND COMPOSITES

Glass fibre is stronger than metals in bulk form by virtue of the fact that its structure contains fewer flaws or defects. Similarly, other fibres and particularly whiskers may exhibit even more impressive strength or stiffness values owing to the perfection of their crystal structures. The use of fibres and whiskers having highly ordered and defect-free structures can provide the key to further material development, but, in order to make them of practical value, they must be joined and held in some way—by a bonding material or matrix. Fibres and whiskers can thus serve to reinforce the matrix, which in turn holds them all in place. This leads to the concept of a composite body having a two-phase structure: a fibre which acts as the reinforcement, and a matrix which bonds and holds all the fibres. To obtain the highest strength and modulus from such a composite, it is necessary that each fibre should take its full share of the load, and that the maximum number of fibres be contained per unit volume since the fibre determines the strength.

Glass fibres have found an ever-increasing use as an engineering structural material, the normal practice being to bond the fibres together by resins chosen to give good adhesion to the fibre surface and to have stress–strain characteristics compatible with the fibres.

**Table 1.1** TYPICAL PROPERTIES OF MATERIALS IN BULK AND FIBRE FORM

| Material | Specific gravity | Ultimate tensile strength (lbf/in²) | | Tensile modulus (10⁶ lbf/in²) | |
|---|---|---|---|---|---|
| | | Actual | Specific | Actual | Specific |
| Steel piano wire | 7·8 | 350000 | 45000 | 30·0 | 3·8 |
| 85 ton aircraft structural steel (85 tonf/in² tensile strength) | 7·8 | 190000 | 24000 | 30·0 | 3·8 |
| Duralumin | 2·8 | 75000 | 25000 | 10·5 | 3·7 |
| Magnesium alloy | 1·8 | 43000 | 24000 | 6·5 | 3·6 |
| Tungsten wire | 19·3 | 550000 | 30000 | 50·0 | 2·6 |
| Wood (oak) | 0·7 | 10000–14000 | 14300–20000 | 1·5–2·0 | 2·0–3·0 |
| E-glass | 2·5 | 400000–500000 | 160000–200000 | 10·5 | 4·2 |
| S-glass | 2·5 | 500000–700000 | 200000–280000 | 12·5 | 5·0 |
| Drawn silica | 2·5 | 860000 | 350000 | 10·5 | 4·2 |

It should be noted that the strengths of freshly drawn glass and silica fibres may be as high as 1 000 000 lbf/in². These materials however are susceptible to surface damage and in practice E-glass, for example, has an ultimate tensile strength nearer to 250000 lbf/in².
The specific strength and modulus figures in the table have been obtained by dividing strength and modulus by the specific gravity of the material: hence the values indicate their relative performance per unit weight.

It is important to note that the mechanical properties of any composite material are generally determined by the quantity of fibre per unit volume and by the properties of the fibre. At the same time, since the fibre is much stronger than the resin matrix, the strength of the composite must be lower than that of the individual fibres. Thus, in a two-component material containing 50% by volume of unidirectionally laid fibre, the strength will be halved in the direction of the fibre alignment compared with the strength of fibres themselves. In other directions, the strength will be even lower. For example, with an E-type glass having an ultimate tensile strength of 400000 lbf/in$^2$, a composite would have a tensile strength in the region of 200000 lbf/in$^2$ if all the filaments (50% by volume) were parallel to the direction of the tensile load. At right angles to the fibres, the strength would depend largely on the strength of bonding between the fibres and the resin.

The majority of structures require strength and stiffness in more than one direction. This is obtained by laying fibres in the required directions in the same plane, as in the case of a composite sheet or plate. The properties in any direction will be governed mainly by the amount of fibres in that direction. Chopped glass fibre is widely used to obtain random or three-dimensional orientation in one composite body; such a method of manufacture is relatively cheap and allows complex shapes to be produced. However, it is difficult to incorporate a high proportion of fibre into the composites, and their properties are correspondingly low.

Nevertheless, glass fibre–resin systems may possess appreciably higher specific strengths than those achievable in conventional materials in bulk form. Their specific stiffness, on the other hand, is only marginally greater*. However, it should be noted that, quite apart from mechanical properties, glass-fibre reinforced resins or plastics (GRP) have other advantages over metals. For example, they are usually easy and relatively cheap to fabricate in a wide variety of shapes and sizes. GRP composites are also characterised by their good chemical inertness towards many substances which would corrode metals. It is not surprising, therefore, that GRP materials have found widespread and steadily increasing use. In fact, composites as a whole may be regarded as forming a new

---

* Part 1 of this monograph is concerned with fibres and their properties. The strength of a fibre refers to its ultimate tensile strength based on axial loading of the fibre. Similarly, the modulus (a measure of stiffness) refers to the tensile modulus (Young's modulus), which is the ratio of the tensile (axial) stress to the tensile (axial) strain. Specific strength and specific modulus are derived by dividing the strength and modulus respectively by the specific gravity of the fibre. Part 2 deals with composites, and frequent reference is made in it to the flexural strength and modulus of composite test bars.

family of engineering materials having considerable growth potential and whose properties may be varied over wide limits as in the case of metal alloys.

Within the last few years, further advances have taken place in the production of other fibres for reinforcement, the major improvement being in the area of specific stiffness. Thus, a new range of composites may be formulated having greatly enhanced properties relative to GRP materials.

## 1.3    FIBRE REINFORCEMENT MATERIALS FOR COMPOSITES

Glass fibre paved the way for reinforced resin structural materials in the late 1930s. Since that time, steady improvements and continuous market growth have take place. Glass fibre continues to be the most widely used reinforcement for plastics because of its high strength, ready availability to known specification, low density, and low cost. However, developments in aircraft have led to a search for better materials for reinforcement, the emphasis being placed on stiffness–weight ratio and ability to withstand higher temperatures than glass.

In the late 1950s, ceramic materials were for the first time produced in the form of whiskers, these consisting essentially of small single crystals virtually free from structural defects and exhibiting exceptionally high strength and stiffness. Although the whiskers were extremely costly to produce, their properties appeared attractive for composites where the ultimate in performance was required. The ceramic materials used up to now are simple compounds of the lighter elements in the periodic table, including aluminium, boron, silicon, and beryllium. Thus, $Al_2O_3$, $BeO$, $SiC$, $SiO_2$, and $B_4C$ can be obtained in the form of whiskers, having stiffness values over five times that of the best glass fibre together with improved tensile strengths. Moreover, they are light in weight and possess good refractory properties. At present, much work aimed at the effective utilisation of whiskers in high-performance composite materials is being carried out, but in general the cost of manufacture is very high and sophisticated methods are needed to incorporate and align the whiskers satisfactorily in a matrix.

The element boron has a high specific strength and stiffness if its structure can be obtained in a highly oriented and defect-free form. It has been possible to achieve this on a production scale by producing the boron from a vapour-phase reaction; in practice, the high-temperature reaction is carefully controlled and a coating of the material builds up on a tungsten wire substrate. The resulting product usually has a large diameter compared with glass fibre and may be regarded as a wire reinforcement. Tensile modulus (stiffness) values of $55 \times 10^6$ lbf/in$^2$ have been obtained, with ultimate tensile strengths up to 400000 lbf/in$^2$. Thus, the material possesses the same strength as glass fibre but five times the stiffness. Since the product can be made in long and continuous lengths, it is particularly attractive as a resin reinforcement. However, its cost is high. More recently, silica has been used as a substrate in an attempt to reduce manufacturing costs, but the process still remains expensive.

Carbon holds an important place in the periodic table. It is one of the lightest elements and also the most refractory, retaining its strength up to 2000°C and above. In the octahedral crystal form (diamond), it is the hardest material known to man, whilst, in the hexagonal form (graphite), carbon is an excellent lubricant and has good chemical inertness towards many reagents. It is possible to produce carbon in a crystal form (whiskers) having an ultimate tensile strength in the range $2 \cdot 5$–$3 \cdot 0 \times 10^6$ lbf/in$^2$, and a tensile modulus as high as $100 \times 10^6$ lbf/in$^2$. Such whiskers are laboratory curiosities at the present time and are totally uneconomic in price, but they indicate the tremendous potential properties offered by this chemical element if it can be produced in a form such that the crystal structure is free from the defects and flaws existing in all bulk materials.

At present, carbon fibre rather than carbon in whisker form is regarded as a most important material for reinforcement of composites. Whereas low-strength and low-stiffness carbon fibre has been well known for many years, being made by careful heating of organic fibres such as cellulose or rayon, new techniques have been developed resulting in improved properties. The concept of manufacturing fibres by the controlled pyrolysis of man-made organic yarns such as rayon is particularly attractive since these materials can be obtained with very consistent properties which contribute to the quality of the final product. It is now possible to make carbon fibre from organic starting materials or precursors with tensile modulus values up to $75 \times 10^6$ lbf/in$^2$ or even higher and with strengths up to 500000 lbf/in$^2$, whilst future developments will certainly bring further increases in performance. Although these properties are much inferior to those of carbon whiskers, outstandingly good resin-bonded composites can be readily made from them. Moreover, the organic precursors are basically low in cost and freely available, implying that, for large-scale production, carbon fibre should be competitive in terms of price for a given performance when compared with other reinforcement materials.

## 1.4   PROPERTIES OF VARIOUS REINFORCEMENT MATERIALS

Figure 1.1 indicates the relative strength and modulus values for a number of structural and reinforcement materials including carbon

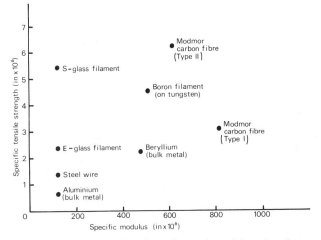

*Figure 1.1. Specific tensile strength and specific tensile modulus values for a number of reinforcing materials: movement towards the top right-hand corner of the graph indicates improved properties; the specific values on this particular graph were obtained by dividing the strength and modulus by the density in lb/in³*

fibre. The values are based on the specific properties since these are particularly relevant for structural materials.

## 1.5   TERMINOLOGY

Reinforcement materials and composites are advancing steadily in terms of both usage and range of applications. It is true to say that a new technology has been established involving a number of specialised processes and techniques. Since those working in the area may have widely differing backgrounds, it is useful at this stage to define the more important terms employed:

*Fibre\**. Any material in an elongated form such that the ratio of its minimum length to its maximum average transverse dimension is 10:1, its maximum cross-sectional area is $7.9 \times 10^{-5}$ in²

\* Suggested by ASTM Committee D30 at Philadelphia, Pennsylvania, on 5 October 1966. Other definitions given in this book are the author's.

B

(corresponding to a circular cross-section of 0·010 in diameter), and its transverse dimension is not greater than 0·010 in.

*Fibre composite material.* A material consisting of two or more distinct physical phases, one of which is a fibrous phase dispersed in a continuous matrix phase. There is no size limitation for the fibrous phase, but it must retain its identity in the composite.

*Filament.* A continuous fibre.

*Matrix.* A bonding material which adheres to and contains the fibres. Many materials, such as the thermoplastic and thermosetting resins, metals, glass, or ceramic materials, can form a matrix. Resins are the most widely used.

*Pre-preg (pre-impregnated fibre).* An intermediate product consisting of fibres or tows which have been coated with a matrix material such as a resin. The fibres are aligned in the majority of cases to give a flat sheet or tape. Usually the resin is not fully cured so that the aggregate remains flexible and the sheet can be built up in plies to form a composite. The technology is described in detail later and includes other forms such as tape, broad goods, and filament winding.

*Whisker\*.* Any material that fits the definition of a fibre and is a single crystal.

*Wire.* A metallic filament.

*Yarn or tow.* A number of filaments in a bundle which can be handled as a single unit. A tow is usually bigger than a yarn, having thousands of filaments, whereas a yarn usually has a few hundred filaments. A yarn may be spun from staple fibre, but a tow is never spun.

## 1.6 OTHER COMPOSITE MATERIALS

A wide range of other composite bodies has already become fully established in everyday use. Examples are: (a) paper combined with phenol- or cresol-type resins to give good strength and electrical insulation properties, and (b) cotton combined with these and other resins to give materials for use in electrical insulation and areas where greater strength is required relative to (a). A further important group of composites is centred on asbestos combined with a wide variety of resins; here the uses are very large, and a number of forming techniques may be employed. One of the

---

\* Suggested by ASTM Committee D30 at Philadelphia, Pennsylvania, on 5 October 1966. Other definitions given in this book are the author's.

cheapest everyday composite materials is hardboard, consisting of wood chippings and glue.

## 1.7   ADVANTAGES OF HIGH–PERFORMANCE COMPOSITE MATERIALS

Composite bodies have the basic advantage that they can be fabricated from a wide choice of reinforcement and matrix materials; their properties may therefore vary over a very considerable range. In general, composite materials possess a number of advantages relative to conventional bulk materials including metals. These are summarised below:

1.  They can be made with high strength and high specific strength relative to general materials of construction.
2.  They can be made with high stiffness and high specific stiffness relative to general materials of construction.
3.  Density is generally low.
4.  Strength can be high at elevated temperatures.
5.  Impact and thermal shock resistance (toughness) are good.
6.  Fatigue strength is good—often better than for metals.
7.  Creep strength is good—often better than for metals.
8.  Oxidation and corrosion resistance are particularly good.
9.  Thermal expansion is low and can be controlled.
10. Thermal conductivity can be controlled, and so can electrical conductivity.
11. Stress–rupture life is improved relative to many metals.
12. Predetermined properties can be produced if required to meet individual needs.
13. Fabrication of large components can often be carried out at lower cost than for metals.

# 2

# Development and uses
# of carbon fibres

## 2.1 LAMPS AND HIGH-TEMPERATURE APPLICATIONS

Carbon fibres and filaments are not new; they were first produced
by Edison[1] almost a century ago for electric lamps*. The method of
manufacture involved the careful carbonisation of cellulose strands
to convert the organic matter into carbon. Bamboo or even cotton
was used as the starting material, and it had to be bent to the
required shape before being placed in the carbonising chamber since
it was extremely brittle when processed and required the most
delicate handling. Later, Edison made consistent filaments by
dissolving good clean commercial cotton in a suitable solvent,
usually a solution of zinc chloride, and squirting the resulting mass
through a die into a fluid hardener. Superior filaments were made
by depositing pyrolytic carbon on to such a base filament by
'flashing' the filament in a hydrocarbon atmosphere. Flashing was
achieved by passing a sufficiently heavy electric current to heat the
filament to a white heat. In 1909, Whitney[2] patented an improved
filement for electric lamps by taking the flashed filaments and

---

* T. A. Edison obtained a US Patent and UK Patent No. 4576/1879 dated 10 November 1879, relating to carbon filaments for electric lamps. On 27 November 1880,
J. W. Swan also took out a UK provisional patent specification for lamp filaments
made out of carbon (No. 4933/1880). These are among the first recorded patents
dealing with carbon fibres. A reproduction of UK Patent No. 4576/1879 is shown in
Figure 2.1.

heating them to a very high temperature in a carbon tube furnace. The resulting filaments had better electrical resistance characteristics. For the electric lamp application, mechanical properties of the filaments were of secondary importance to uniformity of electrical resistance, and the contemporary filaments were of no value whatsoever as reinforcement materials. Moreover, their use

A.D. 1879, 10th *November.* N° 4576.

### Electric Lamps.

LETTERS PATENT to Thomas Alva Edison, of Menlo Park, in the State of New Jersey, United States of America, for the Invention of " IMPROVEMENTS IN ELECTRIC LAMPS, AND IN THE METHOD OF MANUFACTURING THE SAME."
Sealed the 6th February 1880, and dated the 10th November 1879.

PROVISIONAL SPECIFICATION left by the said Thomas Alva Edison at the Office of the Commissioners of Patents on the 10th November 1879.

THOMAS ALVA EDISON, of Menlo Park, in the State of New Jersey, United States of America. " IMPROVEMENTS IN ELECTRIC LAMPS, AND IN THE METHOD OF 5 MANUFACTURING THE SAME."

It is necessary to practically subdivide the electric light into a great number of luminous points, so that lamps connected in multiple may be employed without the necessity of using conductors of great size for the current to the lamps. It is essential that the lamps should be of great resistance. In the Provisional 10 Specification of Patent dated June 17th, 1879, No. 2402, I have set this forth, and obtained lamps of great resistance by employing long lengths of platinum or metallic wires pyroinsulated, and wound in such a manner that but small radiating surface is exposed to the air, although a great length of wire is used.

My present Invention relates to lamps of a similar character, except that carbon 15 threads or strips are used in place of metallic wires. I use a block of glass, into which are sealed two platinum wires. These wires serve to convey the current to the electric lamp within a bulb, which is blown over the lamp and united to the glass block. The bulb is exhausted of air to about one millionth of an atmosphere. Upon the ends of these wires are two clamps that secure two other platinum 20 wires. The burner consists of a filament or thread of carbon, preferably coiled, with the ends secured to the platinum wires, the whole being made as follows :—

Fibrous material, such as paper, thread, wood, or any vegetable or animal matter which can be carbonized, has the ends secured to platinum wires, the fibre is 25 wound in such a shape as to expose the least amount of surface to radiation, such as in a helix or spiral. The helix is secured to the platinum wires by plastic carbon, and the whole is placed in a closed vessel free from air, and subjected to a heat sufficient to fully carbonize the fibre, and leave nothing but carbon. At the

[*Price 6d.*]

*Figure 2.1. Reproduction of original T. A. Edison patent; this corresponds to US Patent No. 223 898, filed 4 November 1879*

was short-lived owing to the advent of the metal-filament in-candescent lamps.

However, the use of carbon fibre has continued for other applications where its refractoriness and chemical inertness have proved valuable. Thus, high-temperature furnaces operating with non-oxidising atmospheres frequently use carbon fibre for insulation, the carbon being in the form of felt, wool, braid, or cloth. The extreme refractoriness of these materials enables such furnaces to operate at temperatures up to 3000°C; moreover, the low bulk density of the fibres reduces the thermal mass of the structure. Graphite textiles are also used in the processing of refractory metals, such as the controlled carburising of vanadium, niobium, and tantalum. A large number of patents dealing with such applications has been taken out over the last 20 years and References 3–15 describe the use of carbon fibre for high-temperature insulation. It is useful to point out the difference between carbon and graphite fibre at this stage. Both are based on the element carbon, but the graphitic form is made by heating the carbon to a high temperature (over 2000°C) so that a graphite-type crystal structure develops. Graphite filaments have a higher electrical conductivity than carbon filaments of the same length and cross-sectional area and tend to have a more slippery surface, characteristic of graphite in bulk form. The lower-temperature carbon fibre has a finer crystal structure, which may be more disordered than in graphite.

## 2.2   CHEMICAL PLANT

For chemical plant where resistance to corrosion is important, carbon has found widespread use, both in bulk form and as felt, wool, braid, or cloth; the applications are packings, glands, seals on valves, pumps, and bearings, and catalyst supports. Graphite fibre packings are particularly effective for the shafts of centrifugal pumps, agitators, churns, driers, and blowers in the chemical industry. The packing can be made relatively leakless in operation, and its thermal conductivity approaches that of cast iron, allowing heat from the shaft–packing interface to be conducted away. Except for strong oxidising agents, chemical resistance is good up to 350°C. In inert or special atmospheres, carbon fibre can operate at much higher temperatures, a notable example being a valve gland packing for superheated steam; in modern electrical generating stations, steam temperatures frequently reach 560°C, conditions under which asbestos packings start to decompose owing to dehydration. Carbon braids and cords can be used in such cases, the only requirement being that air is kept out of contact with the carbon. It should be noted that carbon braids and cords provide some degree of lubrication, which is an advantage. A number of proprietary brands of carbon seals and packing materials is available, and frequently these are combined with thermoplastics such as nylon or PTFE to improve the sealing and lubrication properties. The majority of chemical plant applications demand chemical inertness from the carbon fibre, and hence its chemical purity may be important. As with the furnace applications, strength is not of prime importance but consistency, particularly in terms of flexibility, handling qualities, and freedom from brittleness, is necessary; hence the process of conversion from organic precursor must be carefully controlled. References 16–28 describe many applications of carbon in the chemical industry.

## 2.3 MISCELLANEOUS APPLICATIONS

Other uses include electric heating elements for special environments. For applications where the element has to be extremely flexible, as in an electric blanket, for example, carbon may be employed with advantage, either as braid or cord or in the form of felt. In each case, the carbon material is sandwiched between outer layers of an ordinary wool blanket. Carbon and glass yarns have been used to make a heater in cloth form, a possible application being electrically heated seats for cars. A particularly interesting development is the use of carbon–glass suits (carbon and glass fibre) for the 'hotline' maintenance of transmission lines of up to 765 kV. Where heating elements have to operate under corrosive conditions, those made of metals are prone to deteriorate rapidly but carbon can often be used successsfully. For very high temperature use, it is possible to construct electric furnaces in which the heating element is graphite cord or tape, wound on a refractory former. Graphite cloth has also been used for many applications, including the detecting element in electrical measuring instruments, an example being the continuous detection of acid vapours in a stream of air. References 29–45 deal with the uses of carbon cloths, textiles, etc., in heating elements and special applications. Figure 2.2 shows a range of low-modulus carbon cloths, felts, wools, etc., for a variety of purposes.

The variety of applications described above is not intended to be complete but serves to indicate the usefulness of carbon in filamentary or textile form. The filaments used are not required to possess either high strength or high stiffness in order to be effective.

In fact, their properties are such that they are entirely unsuitable for reinforcing resins in order to obtain a high degree of mechanical strength or stiffness. Nevertheless, their technology has provided a

*Figure 2.2. Example of carbon textile materials including yarn, felt, cloth, braid, tape, and wool. Courtesy Le Carbone (Great Britain) Ltd*

basis for the development of carbon fibres having greatly improved mechanical properties.

## 2.4    ABLATIVE CHARACTERISTICS FOR AEROSPACE APPLICATIONS

In recent years, an entirely new use has emerged for carbon fibre. This has arisen in the aerospace industry, where there is a requirement for materials to resist the enormous temperatures developed on the re-entry of space vehicles into the earth's atmosphere. Normal refractory materials are not capable of withstanding these temperatures and, in particular, the tremendous thermal shock at the point of entry of the vehicle into the earth's atmosphere. Thus, the most successful materials to date are based on carbon fibre and resins. A carbon-fibre reinforced phenolic resin, for example, is particularly effective in resisting these temperatures since the resin evaporates and burns at the surface. Much of the heat due to friction caused by the earth's atmosphere is absorbed in vapourising the resin, and the low thermal conductivity* of the fibre–resin mass prevents its disintegration below the surface. The material thus acts as an ablative substance whereby latent heat of evaporation allows extremely high temperatures to be withstood for short periods. At the present time in the Western world, this application exists almost entirely in the USA, owing to their extensive aerospace programme, in which relatively large quantities of fibre are used. As a rule, very high strength or stiffness is not required and the fibre can be of the same type as for the previous applications. However, it should be noted that, for ablative applications, the stability of the resin is important and the fibre should be as chemically inert as possible; in particular, certain impurities can catalyse breakdown and speed degradation of the composite and must therefore be avoided. Textiles made from graphitic fibres are often more suitable for ablative applications, however, despite their higher thermal conductivity. The reason lies in the purity level; carbon-based textiles often contain significant quantities of alkali metals which originate in the precursor and which catalyse the decomposition and breakdown of the resin system. The latter is frequently a phenolic-type owing to its high coking value and good high-temperature stability.

The ablative properties of carbon-fibre resins have also led to their use for rocket nozzles. Here the temperatures encountered are very high indeed coupled with extremely severe thermal shock. Since a rocket is required to operate for a relatively short period, vapourisation and combustion of the fibre–resin allows the structure to remain intact during this period. Further details are given in References 46–48.

## 2.5   REINFORCEMENT

The most recent application for carbon fibre is for the reinforcement of resins or even metals to yield composite materials of exceptionally high specific stiffness and strength. The fibre for such uses must have greatly increased mechanical properties compared with the grades employed in the previous applications. To achieve these properties, special manufacturing processes have been developed and are described in the next two chapters.

---

* The actual carbon fibre possesses high thermal conductivity as stated in Section 2.2, but the high resin content provides heat insulation between fibres. Moreover, for use in ablative applications, the fibres are aligned parallel to the surface to reduce heat flow from the surface into the composite.

C

REFERENCES

1 EDISON, T. A., US Pat. 470925 (15.3.1892)
2 WHITNEY, W. R., US Pat. 916905 (30.3.09)
3 HIGH TEMPERATURE MATERIALS INC., 'A Pure, Flexible, Insulating Graphite Tape with Highly Directional Properties Similar to Pyrolytic Graphite has been Developed', *Ind. Res., Chicago,* **4** No. 9, 14 (1962)
4 LE CARBONE (GREAT BRITAIN) LTD, 'High-Temperature Insulation: Fibrous Carbon and Graphite', *Mach. Des. Engng,* **2**, 63 (1964)
5 NATIONAL CARBON CO., 'Graphite Textiles for Industry', *Br. chem. Engng,* **6** No. 9, 592 (1961)
6 NATIONAL RUBBER CO., 'Fiber Stretches Graphite Use', *Chem. Engng, Albany,* **66** No. 9, 70 (1959)
7 NATIONAL CARBON CO., 'Graphite—Newest of the "Wonder Fabrics"', *Mach. Des.,* **31** No. 9, 32 (1959)
8 NATIONAL CARBON CO., 'Refractory Fabrics that Conduct Heat and Electricity', *Product Engng,* **30** No. 18, 7 (1959)
9 SUNBEAM EQUIPMENT CORP., 'Electric Furnace Prevents Formation of Compounds', *Am. Mach./Metalwkg Mfg,* **109** No. 23, 142 (1965)
10 H. I. THOMPSON FIBER GLASS CO., *Carbon and Graphite,* Form 393 A.P. (1965); *Hitco-G Graphite Materials,* Technical Bulletin PMD G-100 (1965)
11 H. I. THOMPSON FIBER GLASS CO., 'Graphite Cloth Aids Carburizing', *Steel,* **155** No. 14, 100 (1964)
12 THE BENDIX CORP., US Pat. 3159524 (1.12.64)
13 DEERING MILLIKEN RESEARCH CORP., US Pat. 3281261 (25.10.66)
14 HAVEG INDUSTRIES INC., Brit. Pat. 1005107 (22.9.65)
15 EUROPEAN ATOMIC ENERGY COMMUNITY (EURATOM), Brit. Pat. 1147905 (10.4.69)
16 GARLOCK INC., 'Graphite Filament Packing Reduces Fluid Leakage', *Power,* **110** No. 2, 98 (1966)
17 GARLOCK INC., Advertisement for graphite filament packings, *Chem. Engng, Albany,* **73** No. 8, 83 (1966)
18 THE MARLO CO., 'New Packing Beats Both TFE and Graphite–Fiber Types', *Chem. Engng, Albany,* **75** No. 22, 72 (1968)
19 RAYBESTOS-MANHATTAN INC., 'Graphite Filament Packing', *Mach. Des.,* **40** No. 12, 254 (1968)
20 UNION CARBIDE CORP., *Graphite Products,* Technical Information Bulletins 505-JB, 505-IC, 507-JC, 509-BD (n.d., probably *c.* 1964)
21 NATIONAL CARBON CO., DIVISION OF UNION CARBIDE CORP., 'Fibrous Graphite', *Engng Mater. Des.,* **6** No. 8, 591 (1963)
22 NATIONAL CARBON CO., DIVISION OF UNION CARBIDE CORP., 'National Carbon Packing Material Offered for Valve and Pump Use', *Am. Mach./Metalwkg Mfg,* **107** No. 15, 91 (1963)
23 NATIONAL CARBON CO., DIVISION OF UNION CARBIDE CORP., 'Stronger, More Flexible Graphite Cloth', *Mater. Des. Engng,* **53** No. 5, 7 (1961)
24 NATIONAL CARBON CO., DIVISION OF UNION CARBIDE CORP., 'Graphite Cloth', *Product Engng,* **32** No. 41, 122 (1961)
25 THE WOODVILLE RUBBER CO (FINE PRODUCTS) LTD, 'Packing Material'. *Fact. Equip. News,* **20** No. 257, 10 (1968)
26 SOCIÉTÉ GÉNÉRALE DES PRODUITS RÉFRACTAIRES, Brit. Pat. 1150434 (30.4.69)
27 SOCIÉTÉ LE CARBONE-LORRAINE, US Pat. 3446593 (27.5.69)
28 C. I. HAYES INC., 'Vacuum Furnace', *Semicond. Prod. Solid St. Technol.,* **6** No. 6, 48 (1963)
29 C. I. HAYES INC., Advertisement for high-purity graphite cloth, *Metal Prog.,* **82** No. 2, 48 (1962)

30 HERSCH, P., and SAMBURETTI, C. J., 'Cell for Measuring Acid Vapours in Gases', *R & D*, No. 36, 28 (1964)

31 ROHR AIRCRAFT CORP., 'Graphite Cloth Speeds Heat Cycle', *Steel*, **146** No. 8, 74 (1960)

32 T-M VACUUM PRODUCTS, 'Heat Retainer for 2400°F Furnace is Water Cooled', *Am. Mach./Metalwkg Mfg*, **110** No. 15, 164 (1966)

33 TRI METAL WORKS INC., 'Tri Metal 3000°F Vacuum Furnace', *Ind. Heat.*, **32** No. 4, 761 (1965)

34 UNION CARBIDE CORP., 'Electrically Conducting Cloth', *Automot. Inds*, **133** No. 10, 108 (1965)

35 UNION CARBIDE CORP., 'No Waiting' (car seat heaters), *Financial Times*, No. 23761, 13 (29.10.65)

36 OWENS-CORNING FIBERGLAS CO., 'Carbon Glass Suits Resist High Voltages', *Ceramic Ind.*, **89** No. 1, 38 (1967)

37 UNION CARBIDE CORP., Technical Bulletins 101 DA, 104 LJ, 105 CB (n.d.)

38 BAIRD-ATOMIC INC., US Pat. 3 120 597 (4.2.65)

39 BAIRD-ATOMIC INC., US Pat. 3 236 205 (22.2.66)

40 GREAT LAKES CARBON CORP., US Pat. 3 399 266 (27.8.68)

41 AIR REDUCTION CO. INC., US Pat. 3 400 253 (3.9.68)

42 BAIRD-ATOMIC INC., Brit. Pat. 996 338 (23.6.65)

43 C. I. HAYES INC., Brit. Pat. 1 124 151 (21.8.68)

44 BUITKUS, T. J., US Pat. 3 174 880 (23.3.65)

45 J. P. STEVENS & CO. INC., 'A New Carbon Fabric will Have Aerospace Applications', *Chem. Engng News*, **44** No. 47, 47 (1966)

46 UNION CARBIDE CORP., 'Carbon Cloth', *Chem. Engng, Albany*, **72** No. 23, 122 (1965)

47 NATIONAL CARBON CO., 'Stronger Graphite Cloth', *Iron Age*, **187** No. 13, 105 (1961)

48 WESTINGHOUSE ELECTRIC CORP., 'New Laminates for Heat Resistance', *Mater. Des. Engng*, **55** No. 4, 10 (1962)

# 3

# Production of carbon fibres from cellulose and other organic materials excluding polyacrylonitrile

## 3.1 PRODUCTION OF LOW-STRENGTH AND LOW-MODULUS FIBRE FROM CELLULOSE AND OTHER MATERIALS

### 3.1.1 STRUCTURAL CHANGES ON DECOMPOSITION OF CELLULOSE

Cellulose is an organic compound of the polysaccharide type having the formula $(C_6H_{10}O_5)_n$. It is the substance which composes the greater part of the cell walls of plants: cotton fibres, for instance, consist of over 90% pure cellulose. It undergoes thermal decomposition without the presence of a melting stage, forming a strong carbonaceous residue. Fibrous cellulose can therefore be used for the production of carbon fibres and textiles for the applications described in the previous chapter, and Cranch[1] describes the production of carbon fibrous materials from cellulose. However, the conversion of cellulose to carbon by heating in an inert atmosphere results in a very low yield of carbon material, of the order of 15% or even lower. But the yield can be increased up to 25% or 30% by very slow heating of the cellulose, a fact which indicates that the chemical changes taking place are by no means simple.

Tamaru[2-4] and Parks et al.[5] state that the pyrolytic decomposition of cellulose involves depolymerisation at elevated temperatures through cutting or breaking of the 1,4 glycosidic linkage followed by intramolecular rearrangement of the cellulosic units to give laevoglucosan. Study by Schwenker and Pascu[6] of the pyrolytic degradation of untreated cellulose and its derivatives revealed that heating in air at 350°–375°C produced laevoglucosan as an inter-mediate product and that it underwent fragmentation to form volatiles and low molecular weight products. In addition, polymeri-sation and aromatisation occurred with the formation of chars. Later work by Madorsky, Hart, and Straus[7,8] involved the pyrolysis of cellulose under vacuum or inert atmospheres; random breaking of C—O bonds and simultaneous dehydration were suggested for the thermal breakdown. Infra-red absorption spectroscopy was used by Hofman et al.[9] to examine the decomposition of cellulose up to 575°C, and the work showed the value of infra-red techniques for monitoring structural changes.

Tang and Bacon[10] carried out much useful work in this field. They showed that there is a marked weight loss and a length shrinkage in the temperature range 240°–320°C for all samples examined. The rise of the weight loss and shrinkage curves is steep (Figure 3.1), the

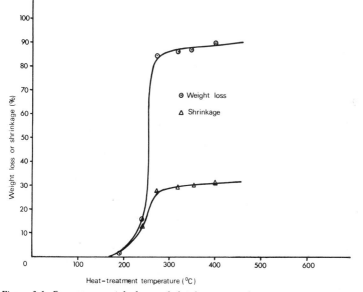

Figure 3.1. Percentage weight loss and shrinkage versus heat-treatment temperature during pyrolysis of Fortisan 36 cellulose fibre, which is produced by American Celanese Corp.; samples were heated at 40°C/h in a slow flow of argon. From Tang and Bacon[10], courtesy Maxwell Scientific International, Inc.

latter lagging behind the weight loss and indicating that the major pyrolytic breakdown occurs in the range 240°–320°C. Tang and Bacon refer to the work of Madorsky, Hart, and Straus[7,8] and of Major[11], who state that aqueous distillates, tars, and gases (CO, $CO_2$, $H_2$, etc.) are produced by thermal decomposition of cellulose, the relative amounts depending on both the pyrolysis conditions and the starting material. Thus, at low temperatures up to 250°C, water is the major product; at higher temperatures, much tar is produced, while cotton tends to yield more tar than viscose rayon. Madorsky and his colleagues[7,8] analysed the tar and showed the presence of hydroxyl, methylene, methyl, carbonyl, and olefinic groups. Other workers[6,12,13] have reported a large range of compounds in the tar, some of which are complex. Tang and Bacon carried out infra-red absorption spectroscopy on cellulose degradation products and postulated a multi-stage mechanism for the conversion of cellulose to carbon:

Stage I      Physical desorption of water (25°–150°C)
Stage II     Dehydration from the cellulose unit (150°–240°C)
Stage III    Thermal cleavage of the cyclosidic linkage and scission of other C—O bonds and some C—C bonds via a free radical reaction (240°–400°C)
Stage IV     Aromatisation (400°C) and above

Figure 3.2 shows the proposed mechanism for the conversion of cellulose to carbon. Tang and Bacon also showed that a relationship exists between the structure of carbonised fibres and that of the raw material employed in their preparation, as indicated by x-ray diffraction, electron transmission microscopy, and measurements of weight and length changes resulting from carbonisation. Their x-ray orientation studies showed a direct correspondence between cellulose molecular orientation and preferred orientation of the carbonised fibre; furthermore, the shrinkage of fibre length during carbonisation decreased with increasing cellulose molecular orientation. Tang and Bacon indicate that the building of a carbon structure probably begins with the formation of 'carbon chains' along the paths of the original cellulose chains, preserving a 'replica' of the original fibre structure.

Figure 3.3 indicates the crystal structures of cellulose I (cotton or natural cellulose), cellulose II (mercerised cotton, viscose rayon, acetate rayon), and graphite. Figure 3.4 shows the geometry of the 'longitudinal polymerisation process' of cellulose ring unit residues into graphite layers. It will be seen that the graphitised length is 83% of the raw fibre length, assuming perfect orientation. In Figure

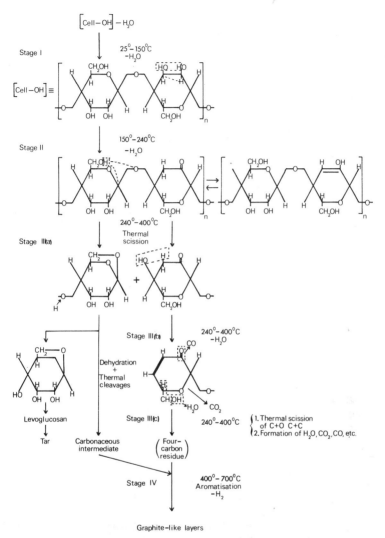

*Figure 3.2. Proposed mechanism for the conversion of cellulose to carbon. From Tang and Bacon[10], courtesy Maxwell Scientific International Inc.*

3.5, which indicates the geometry of the 'transverse polymerisation process', the graphitised length is only 48% of the raw fibre length. Experiments have confirmed that the mechanism given in Figure 3.2 is in agreement with actual results, allowing for the fact that some of the carbon remains in the tar products.

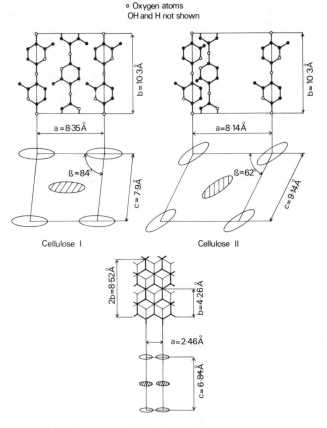

Figure 3.3. Crystal structures of cellulose I, cellulose II, and graphite. From Tang and Bacon[10], courtesy Maxwell Scientific International, Inc.

It will be seen from the above description of the more important structural changes which take place on the decomposition of cellulose that the reactions are very complex indeed. The mode of breakdown of the cellulosic structure depends not only on the starting material—cotton or rayon, for example—but on such factors as heating rates, furnace atmospheres, rate of removal of volatiles, and humidity. In the next section, some guidance is given concerning heating rates and the effect of atmosphere, but it must be remembered that the production of fibres of known and consistent

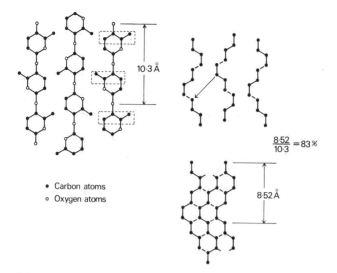

*Figure 3.4. Geometry of the 'longitudinal polymerisation process' of cellulose ring unit residues into graphite layers. From Tang and Bacon[10], courtesy Maxwell Scientific International, Inc.*

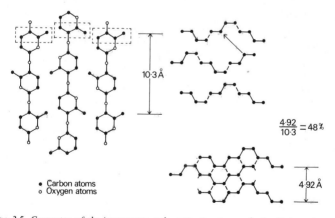

*Figure 3.5. Geometry of the 'transverse polymerisation process' of cellulose ring unit residues into graphite layers. From Tang and Bacon[10], courtesy Maxwell Scientific International, Inc.*

properties requires careful and experienced control of the process coupled with uniformity of the starting material.

### 3.1.2   HEAT-TREATMENT PROCESS FOR THE CONVERSION OF CELLULOSE INTO CARBON FIBRES OR TEXTILES

The first part of this chapter has indicated that up to 90% weight loss occurs when cellulose is converted to carbon and that slow heating rates are necessary if good properties are to be obtained with a high yield in terms of weight of carbon fibre. Ford and Mitchell[14] of the Union Carbide Corp. describe heating conditions suitable for making graphite fibres and textiles, the process being carried out in a protective atmosphere:

1. Temperature rise of from 10°C/h to 50°C/h in the range 100°–400°C.
2. Temperature rise up to 100°C/h in the range 400°–900°C.
3. Heating to about 3000°C until graphitisation takes place.

The strengths of monofilaments are shown in *Table 3.1*, from a table of properties given by Ford and Mitchell. It will be seen that, as the filament diameter is reduced, the tensile strength is increased. This is probably because of the greater ease with which volatiles can be removed from the thinner fibre during the pyrolysis, resulting in a more perfect structure.

The heat-treatment atmosphere can play an important part in determining the mode of breakdown of the cellulose and the final structure obtained—and consequently the properties. Inert atmospheres such as nitrogen demand slow heating rates if good yields are to be obtained. Work carried out by Shindo, Nakanishi, and Soma[15] showed that non-oxidising acid vapours such as hydrogen chloride allow faster heating rates and greater yields (up to 40%) to be obtained (Figure 3.6). These halides act as a catalyst and assist the conversion to carbon. Metal halides may also be used, for example zinc chloride or aluminium chloride; the former is used in the manufacture of active carbon fibre from cellulosic materials[16].

**Table 3.1** STRENGTH OF CARBON FILAMENTS. After Ford and Mitchell[14]

| Fibre origin | Diameter (μm) | Ultimate tensile strength (lbf/in²) |
|---|---|---|
| Ford and Mitchell | 5–7 | 108 000–130 000 |
| Ford and Mitchell | 20–25 | 48 000–53 000 |
| Commercial lamp filaments | ⎰ 146 | 28 000–33 000 |
|  | ⎱ 205 | 18 500–22 000 |

However, to maintain a flexible carbon fibre after conversion, the halide must be in the form of a vapour, and hydrogen chloride is very effective according to Shindo, Nakanishi, and Soma, who report a stepwise breakdown of the cellulose which corresponds to the inflections on their curve of weight loss versus temperature. Zinc and aluminium chlorides are also volatile, and it is interesting to note that both have been used as flame retardants for rayon-type fabrics.

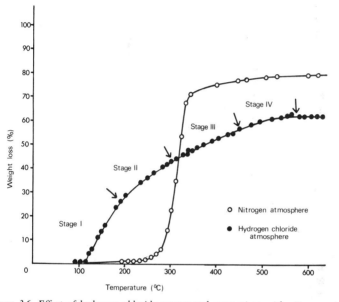

*Figure 3.6. Effect of hydrogen chloride vapour and comparison with nitrogen atmosphere on the weight loss of cellulose (rayon) fibre; rate of temperature rise was 60°C/h. From Shindo, Nakanishi, and Soma[15], courtesy Interscience Publishers.*

A considerable amount of published work exists on methods of making carbon fibres and textiles from cellulose or rayon precursor materials, and many patents have been taken out, some of which describe variations of the heating conditions. It is possible, for example, to heat cellulose-type fibres by placing them in contact with liquids, as opposed to inert gases, in order to facilitate greater heat transfer to the fibre during heating or to impart some particular property to be product[17]. At the end of this chapter is given a list of patents which describe the many variations in processing techniques and the production of fibres and textile materials with specific properties.

### 3.1.3   PRODUCTION OF CARBON FIBRE FROM PRECURSORS OTHER THAN CELLULOSE AND POLYACRYLONITRILE

Many starting materials have been investigated for the production of carbon fibres and textile materials. The aim of much of this work has been to find a starting material which would yield a greater weight of final carbon or graphite product relative to cellulose, of which up to 85% by weight is lost during the pyrolysis, resulting in and accompanied by very considerable shrinkage. An additional aim has been the development of a process which would allow much faster heating rates to be achieved during pyrolysis coupled with the appropriate savings in conversion costs.

Materials such as pitch, asphalt, wool, or lignin have been successfully converted into carbon fibres. All polymers will yield a carbon end-product after decomposition, but, for the production of fibre, the polymer must decompose without melting. Carbon fibres have been made successfully from polymers such as polyvinyl chloride, polyvinyl alcohol, polybenzimidazole, or polyimide; in addition, certain kinds of thermosetting resins based on phenolic groups may be used. With some of these polymers, however, special processing techniques must be employed.

Shindo, Nakanishi, and Soma[18] describe work carried out on polyvinyl alcohol (PVA) or its esters which were first heated in air or halogen to under 400°C and then gradually to 1000°C. For example, 10 g of PVA fibres of 2 denier material were heated for 26 h at 200°C in air, which changed the colour of the material to a black-brown. The fibres were next placed in a quartz tube and heated in nitrogen at 10°C/h to 700°C and then at 50°C/h to 1000°C: 3 g of the carbonised product were then placed in a graphite crucible and heated in nitrogen to 2500°C for 2 h in an electric furnace. The fibres had fairly high strength (75400 lbf/in$^2$) and could be woven into cloth. The work of Shindo and his colleagues indicates that PVA gives low yields in terms of final product weight and still requires slow heating rates. Union Carbide[19] have carried out work on PVA-based fibres in which an oxidation stage is used to prevent the PVA softening so that further heat-treatment stages can be carried out at temperatures in excess of 700°C. No evidence is available to indicate that rapid heating rates are possible. PVA does not therefore appear to offer any exciting prospects for the manufacture of carbon fibres.

Ezekiel[20] carried out experimental work on the conversion of an aromatic polyamide yarn (nylon) into graphite yarn. As in the work of Shindo and his colleagues, the first stage of the conversion involved air oxidation to stabilise the polymer, probably as a result of some form of crosslinking. The carbonisation stage involved pulling the oxidised yarn through a three-stage furnace containing a nitrogen atmosphere. Each stage was about 6 in long, and, during the passage through these, the yarn underwent shrinkage of up to 18% in length. Graphitisation was carried out in a small induction-heated furnace. Tables of properties were given for the fibres, but testing was difficult owing to their brittle nature. Ezekiel concluded that, although the method showed promise for the production of carbon fibre, a major drawback appeared to be the hollow nature of the filaments after graphitising. The majority of carbon fibres obtained by Ezekiel were in the range from $0.7 \times 10^{-3}$ in to $0.9 \times 10^{-3}$ in in diameter. It is likely that, where the volatiles were removed rapidly during the carbonisation stage, excessive pressure build-up occurred within the fibre, opening up the structure and creating internal porosity. The degree of damage caused by volatile removal must increase with fibre diameter because of: (a) the larger quantity of volatiles being liberated, and (b) the greater distance through which the volatiles must diffuse before reaching the surface of the fibre.

Otani, Yokoyama, and Nukui[21] at Gumma University studied the effect of heat treating polyvinyl chloride (PVC) pitch fibre while it was under tensile stress. The aim of the work was to improve the mechanical properties of the fibre by application of tension. Some improvements in strength and modulus were obtained, but the authors concluded that, even after graphitising at 2600°C under stress, the fibre structure was largely isotropic, which would explain the low modulus values. The experience of these workers indicates the difficulty of obtaining the optimum graphite structure in the end-product.

At this stage it must be pointed out that a high modulus of elasticity or high stiffness of the carbon fibre results from a well-developed graphitic structure. Fibre which has not been graphitised possesses a less oriented crystal form and exhibits low modulus values.

Araki and Gomi[22] describe a process for the production of carbon fibre from molten PVC pitch. They indicate that the process has a promising future for the manufacture of carbon fibre owing to the low cost of the starting material and the simplicity of the heat-treatment stages. However, the strength and modulus values of the end-product are not likely to be any higher than for fibres based on other starting materials, notably rayon or cellulose.

Commercial development of the work by Otani and his colleagues has been carried out by a number of Japanese companies including Kureha Chemical Co., Nippon Kayaku Co., and Nippon Carbon Co. In the 'pitch' process, a number of different pitches or lignins may be used as starting materials. Petroleum and PVC pitch have been extensively considered in view of their ready availability and consistent properties. The technique is to melt-spin the pitch into fibres after it has been converted into a suitable physical form. Fibres produced by this method may then be oxidised on the surface to provide hardening and prevent melting during subsequent heat-treatment stages which are necessary to carbonise the organic matter and give a pure carbon fibre. The carbonised material will have an amorphous structure, but crystal development may be imparted to the fibre by graphitisation if required. Many Japanese patents have been taken out by Professor Otani, and some of his additional published work is given in References 23–26. It should be noted that further patents exist based on the work carried out by the Japanese companies themselves, some of this work having been extended to include the production of high-modulus fibre described later in this chapter. At the time of writing, the PVC pitch process promises to be one of the most successful for the production of low-modulus carbon fibres and textiles.

Phenol–hexamine polymers may be extruded from the melt to produce fibre which can be carbonised to form a fine high-strength glassy carbon fibre with an ultimate tensile strength of up to 290 000 lbf/in$^2$ after heat treatment at 900°C, according to Kawamura and Jenkins[27]. Fibres produced by this method have a tensile modulus of $7.12 \times 10^6$ lbf/in$^2$, indicating a more conventional carbon textile material. The strength and modulus increased rapidly with a decrease of fibre diameter, and the glassy fibre had good resistance to chemical attack and could be cheap if produced in quantity from coal tar fractions. Coal Industries (Patents) Ltd[28] have carried out work on fibres from a mixture containing 80–90 % of phenol, the balance consisting of hexamine and the material being extruded to produce fibres. The fibres are carbonised in two stages, first by raising the temperature at 1–2°C/min to 100°–350°C, and secondly by raising the temperature from 350°C to 1250°C followed by final graphitising if required.

Adams et al.[29] report work carried out on the pyrolysis of Saran fibres, which consist of vinylidene chloride–vinyl chloride copolymer, indicating that carbon fibres can be produced but the mechanical properties are not particularly high. The Carborundum Co.[30] reported the production of carbon fibres made by melting novolak resin and drawing fibres from the resin melt; these are then hardened and carbonised, and additives may be included in the resin to improve fibre formation. Again the properties are much lower than those required for high-modulus reinforcing fibres.

It is clear that many starting materials other than cellulose or rayon may be used to give carbon fibres having a modulus of

$4–10 \times 10^6$ lbf/in$^2$ and strengths up to 100000 lbf/in$^2$. Some of these starting materials or precursors are as cheap as cellulose, but they offer an advantage in that their conversion to carbon does not result in such a large weight loss; moreover, their shrinkage is low compared with cellulose. PVC pitch is particularly noteworthy as a starting material, and much development is currently centred on it. However, there does not appear to be any great advantage in the use of these materials relative to cellulose for the production of carbon fibre of very high strength and stiffness. In Section 3.2, consideration is given to the requirements for producing higher modulus and higher strength carbon and graphite fibres from cellulose. These materials are eminently suitable for the reinforcement of plastics and metals.

### 3.1.4    PRODUCTION OF CARBON AND GRAPHITE TEXTILES

The work described in Sections 3.1.1–3.1.3 refers mainly to the processing of single fibres and yarns. These have limited application, and, in practice, the carbon or graphite filaments are often required in a woven form. For some purposes a braid or cord is required, whilst for other uses a felt or low-density wool is needed. For many applications involving insulation a cloth is desirable, and special woven forms are used in ablative-type structures. Thus, the filaments of carbon or graphite are required in some textile form rather than as discrete filaments. Since carbon or graphite fibre cannot be woven easily into braid or cloth owing to its brittleness and difficulty of handling, it is normal practice to start with the organic precursor in the woven form prior to the actual heat treatment.

Considerable quantities of carbon and graphite textiles have been produced in the USA for aerospace programmes. The majority of these textiles were manufactured from rayon precursor in the form of a 1650/720 two-ply yarn converted into the required weave, which normally had a width of 42 in or 48 in and lengths up to 100 ft. For the manufacture of such a material, considerable expertise is required in the heat-treatment stages since all parts of the fabric must undergo the same treatment conditions if a uniform product is to be obtained.

Two types of textile fabrics are normally produced, one consisting of carbon filaments, the other of graphite. In the former, the filaments have a fine carbon-type structure in which the crystal form is not too clearly defined, whilst their purity lies in the range 90–98 % carbon. The graphite textiles, on the other hand, have a graphitic structure and a higher density, whilst their chemical purity is much higher with a graphitic carbon content of 99·9 %, extending as high as 99·99 %. Carbon textiles have a much lower thermal and electrical conductivity than their graphite counterparts and find use in thermal insulation applications and for some ablative uses where the low heat conductivity is an advantage.

In the manufacture of special braids and cords, it is often necessary to incorporate other materials in order to obtain improved sealing or lubrication characteristics. Thus, in the production of braids and gaskets for seals, nylon or PTFE is frequently added, either as an internal core or as a coating on the actual carbon fibres forming the braid. Clearly, for this class of product, the weaving must be carried out *after* the heat-treatment stages, and special machines have been developed to weave both carbon and graphite yarns into the required form, the plastic core being combined at the same time. A range of carbon and graphite textile materials intended for insulation, seals, and packing applications is shown in Figure 2.2.

### 3.1.5    PRODUCTION OF LOW-STRENGTH AND LOW-MODULUS FIBRE—SUMMARY

It will be useful at this stage to draw some conclusions from the extensive investigations reported in this chapter.

Cellulose or rayon is the most widely used precursor or starting material for making carbon fibres. The chemical decomposition is very complex and not fully understood. Cellulose yields 15–30% by weight of carbon fibre and does not melt during decomposition; hence the physical form of the precursor is maintained. However, much shrinkage accompanies the breakdown of the organic structure.

The breakdown of cellulose can be modified by varying the heating rates used, whilst the furnace atmosphere and its moisture content are also important. The type of cellulose used has an influence on the end-product obtained. Additions of catalysts can alter the mode of breakdown, and those containing halides are particularly noteworthy in that they allow faster heating rates to be achieved and increase the carbon fibre yield up to 40%.

The temperature range 250°–300°C is very critical in the decomposition process as the bulk of the weight loss takes place within this temperature range. A large shrinkage accompanies the release of volatiles, and the rate of release must be carefully controlled for optimum results. Research has shown that small-diameter filaments are not as easily damaged as large ones during the volatile removal stage and possess higher strengths when formed under comparable heating conditions. This is because of the greater ease of gas diffusion throughout the fibre.

The structure of low-modulus fibre made from cellulose is crystalline, but the crystals are not well developed and tend to be unoriented. It is the lack of crystal orientation which accounts for the low modulus values obtained.

Whereas all polymers decompose to give carbon, only a few meet the requirement that they should not melt during decomposition and can be used as precursors for carbon fibre manufacture. Much effort has been made to find a non-cellulose precursor which decomposes to give a greater carbon fibre yield with reduced shrinkage. Although progress has been made, the improvements are small and the properties of the end-product are only marginally better. Moreover, it is frequently necessary to go through difficult process stages to convert the polymer into a suitable precursor.

## 3.2 PRODUCTION OF HIGH-MODULUS GRAPHITE FILAMENTS AND YARNS FROM CELLULOSE PRECURSOR

### 3.2.1 STRUCTURE CONSIDERATIONS

The materials described so far in this chapter have a tensile modulus between $4 \times 10^6$ lbf/in$^2$ and $10 \times 10^6$ lbf/in$^2$. Individual filament strengths may be as high as 100 000 lbf/in$^2$. It will be seen that such materials are barely equal to glass fibre in terms of stiffness and possess only about 20% of its strength. For purposes of reinforcement, much higher modulus and strength values are required.

X-ray studies of low-modulus carbon fibre produced from rayon have revealed that the graphite layers tend to be preferentially oriented parallel to the fibre axis. Electron microscopy also confirms these findings and suggests that the orientation is largely derived from the preferred orientation present in the rayon precursor fibre. The important point is that these graphite layers are by no means fully oriented parallel to the fibre axis; moreover, there is evidence of an appreciable degree of microporosity between the individual graphite layers. Estimates from electron microscopy indicate that the pores measure up to 100Å in diameter, whilst density measurements confirm that the fibres possess a much lower density than the theoretical value for graphite.

To obtain greater stiffness and strength, the fibre structure must be improved. Kotlensky, Titus, and Martens[31] studied a pyrolytic graphite deposited from methane gas at 2110°C on to a graphite substrate. The pyrolytic material as deposited had a very high density, around 2·20 g/cm$^3$, which is close to the theoretical density.

In spite of the high density, the tensile modulus was only $2.5 \times 10^6$ lbf/in$^2$. However, when this graphite material was hot worked at 2750°C, a process involving the application of a tensile stress in the basal plane direction, the stiffness improved. After 7% elongation, the stiffness increased to $7.9 \times 10^6$ lbf/in$^2$ and, after 16% elongation, it had increased to the extremely high value of $81.2 \times 10^6$ lbf/in$^2$ which is not very far short of the value of $100-145 \times 10^6$ lbf/in$^2$ obtained by Bacon for graphite whiskers. Thus, the stiffness of the material is very much dependent on the orientation of the individual graphite planes; random orientation results in low stiffness, whilst alignment parallel to the tension axis implies high stiffness. A high degree of orientation is therefore essential in carbon fibre intended for reinforcement purposes.

### 3.2.2    PROPERTIES OF FIBRE MADE BY HOT STRETCHING

The vital step in the production of improved carbon fibre from rayon precursors is to hot work the fibre by stretching. This may be carried out during the graphitising stage, the operation aligning the graphite planes parallel to the fibre axis. At the same time, the process of alignment reduces the internal pore size and brings the individual graphite planes closer together. In effect, the fibre diameter is reduced by this hot-stretching method, and the density is increased. Bacon and Smith[32] carried out experimental work in which carbonised rayon filaments were stretched at high temperature; an elongation of $3.6\%$ gave a modulus increase of 19%, indicating that the structure of a carbon filament may be modified and improved by hot stretching.

More recently, Bacon and Schalamon[33] describe their work on the stretching of carbon fibres at high temperatures to yield a material possessing a high degree of preferred orientation of the graphite layers parallel to the fibre axis. As a result, the fibre tensile modulus was increased from about $6 \times 10^6$ lbf/in$^2$ to $50 \times 10^6$ lbf/in$^2$ and the strength of the filaments was increased considerably, from about 100 000 lbf/in$^2$ to as high as 400 000 lbf/in$^2$. Moreover, it was shown that further stretching could be carried out on the filaments to yield even higher modulus and strength values. In general, experience in the stretching of rayon-based carbon fibres at graphitising temperatures has shown that, as the stiffness increases, so does the strength, owing to the formation of a better structural alignment of the graphite layers and the progressive elimination of porosity. Bacon and Schalamon[33] give data showing the relation between the degree of orientation of the graphite layers and the modulus (see *Table 3.2*).

**Table 3.2** RELATIONSHIP BETWEEN MODULUS AND CRYSTAL ORIENTATION FOR HOT-STRETCHED RAYON-BASED GRAPHITE FIBRES

| Tensile modulus ($10^6$ lbf/in$^2$) | X-ray orientation 'half-width' (degrees) | Density (g/cm$^3$) |
|---|---|---|
| 10 | 25 | 1·35 |
| 40 | 10 | — |
| 80 | 5 | 1·95 |

It will be seen that, as the graphite crystals become closely aligned to the fibre axis, the modulus increases sharply. The density of the fibres also increases as the porosity is suppressed. Bacon and Schalamon[33] found that, since the tensile stress–strain curves are perfectly linear at room temperature, the ratio of tensile strength to tensile modulus is equal to the elastic strain at fracture. This amounts to over $1\%$ for low-modulus fibres and approximately $\frac{1}{2}\%$ for high-modulus fibres. Ruland[34] carried out an analysis of the relationship between the preferred orientation and the tensile modulus of Thornel carbon fibre produced by Union Carbide; this analysis is related to the structure of the graphite layers in the filaments. Hugo, Phillips, and Roberts[35] have also carried out work on the structure of Thornel 50 fibre, and their experiments showed that there are no transverse grain boundaries; instead, the fibre planes were found to 'flow' over long distances in the direction of the fibre axis and to coalesce to form graphite crystallite regions having the size and shape already suggested by x-ray diffraction techniques. It is believed that the planes are capable of flowing around intruding material and possibly flaws and voids, a factor which contributes to the production of consistent high-strength and high-stiffness filaments.

### 3.2.3 HOT-STRETCHING TECHNIQUES

Ezekiel and Spain[36] describe techniques for the stretching of carbon fibre during graphitising at up to 2900°C. Two methods were used, namely passing the fibre through a resistance-heated graphite tube, or heating the fibre directly by the passage of electricity. Stretching was achieved by the use of differential unwinding and rewinding speeds, together with pulleys and suspended weights so that controlled tension could be applied to the fibre during the graphitisation stage. Ezekiel and Spain give a table of properties for the stretched fibre; in general, the tensile modulus and strength increase together as the stretching takes place. The fibre properties are based on tests of from eight to ten filaments from any one sample, for which the average value is given in most cases. The method of testing was based on the work described by Schulman[37]. In their paper, Ezekiel and Spain show photomicrographs of the cross-section of rayon-based graphite fibre, which is uneven and angular. This appearance is typical of fibre produced by the pyrolysis of rayon.

Gibson and Langlois[38] describe a useful and also novel processing technique developed by H. I. Thompson Inc. for the production of Hitco fibre from rayon-based yarn. The main feature of the process is that the fibres are made to carry an electric current which heats them to the desired temperature as they pass continuously between rollers. Graphitising temperatures can be reached by passing sufficient current through the fibres using graphite rollers. Application of a differential speed to the rollers allows the fibre to be stretched at will in order to improve the orientation of the graphite planes. This method is similar to that carried out by Bacon and Schalamon, except that the latter workers used radiant heating. Gibson and Langlois present fibre property data showing that, as the graphitising temperature is increased, higher modulus values are obtained; the strength also increases with modulus. At 2800°C, up to 50% elongation is possible without fibre breakage.

The US Air Force[39] have developed a technique using a glow discharge tube so that the fibre itself becomes electrically energised and acts as a primary electrode. A secondary electrode surrounds the fibre, and the atmosphere in the tube acts as a conductor and allows the fibre to be heated to high temperature if required. If necessary, the fibre may also be subjected to tension in order to provide stretching during the graphitisation stage.

At the time of writing, Union Carbide are in regular production of Thornel fibre intended specifically for the reinforcement of resins, etc., the main grades being Thornel 50 and 75, which have tensile modulus values of $50 \times 10^6 \text{lbf/in}^2$ and $75 \times 10^6 \text{lbf/in}^2$ respectively. The other US manufacturer is H. I. Thompson Inc., who produce Hitco fibre having comparable properties.

### 3.2.4   PRODUCTION OF HIGH-STRENGTH AND HIGH-MODULUS FIBRE—SUMMARY

The work reported in this chapter has shown that strength and modulus are a function of the crystal structure of the fibre. A pure graphite crystal having a high degree of structural perfection has an exceptionally high modulus—over $100 \times 10^6 \text{ lbf/in}^2$—but in one plane only. To achieve similar properties in carbon fibre, it must have a graphite structure and the graphite crystals must be oriented parallel to the fibre axis, or as near parallel as possible. Since the structure of carbon fibre produced from cellulose—or indeed from most other precursors—fails to achieve this degree of orientation, it will have a lower modulus.

To increase modulus values, the orientation must be improved by hot working the fibre: this is achieved by stretching during the graphitisation stage. The crystals are made to slide over each other, and their alignment is improved; the greater the stretching, the better the alignment parallel to the axis, resulting in increased modulus and strength. Hot stretching also reduces porosity and gives higher density values.

A number of interesting techniques has been developed to facilitate the hot stretching. The main difficulty is to control the amount of stretching which takes place; merely applying a fixed tension is useless since the fibre will start to stretch and its cross-section will begin to be reduced. The fibre therefore becomes weaker and will stretch more until it finally breaks. Any device used must apply a controlled amount of stretch if consistent fibre properties are to be obtained.

## 3.3    LISTS OF PATENTS

### 3.3.1    PATENTS RELATING TO CELLULOSE-BASED AND RAYON-BASED LOW-MODULUS CARBON FIBRE AND TEXTILE MATERIALS

| | |
|---|---|
| Brit. Pat. 1 016 351 | US Pat. 3 011 981 |
| Brit. Pat. 1 033 009 | US Pat. 3 053 775 |
| Brit. Pat. 1 034 542 | US Pat. 3 107 152 |
| Brit. Pat. 1 034 666 | US Pat. 3 121 050 |
| Brit. Pat. 1 062 887 | US Pat. 3 174 895 |
| Brit. Pat. 1 071 304 | US Pat. 3 174 947 |
| Brit. Pat. 1 110 596 | US Pat. 3 179 605 |
| Brit. Pat. 1 133 718 | US Pat. 3 235 323 |
| | US Pat. 3 313 596 |
| Fr. Pat. 1 274 825 | US Pat. 3 313 597 |
| Fr. Pat. 1 406 529 | US Pat. 3 337 301 |
| Fr. Pat. 1 486 761 | US Pat. 3 351 484 |
| Fr. Pat. 1 546 067 | US Pat. 3 374 102 |
| Germ. Pat. 2 272 801 | US Pat. 3 413 094 |

### 3.3.2    PATENTS RELATING TO HIGH-MODULUS CARBON FIBRE PRODUCED FROM CELLULOSE

Brit. Pat. 1 157 302
US Pat. 3 454 362
Fr. Pat. 1 551 516

### 3.3.3    PATENTS RELATING TO CARBON FIBRE PRODUCED FROM OTHER PRECURSORS EXCLUDING POLYACRYLONITRILE

Brit. Pat. 1 071 400
Brit. Pat. 1 091 890
Fr. Pat. 1 465 030
Germ. Pat. 1 246 510

#### REFERENCES

1  CRANCH, G. E., 'Unique Properties of Flexible Carbon Fibres', *Proc. 5th Conf. Carbon, Vol. 1*, Pergamon, New York, 589 (1962)

2  TAMARU, K., *J. chem. Soc. Japan, Pure Chem. Section*, **69** No. 1–3, 20 (1948)

3  TAMARU, K., *J. chem. Soc. Japan, Pure Chem. Section*, **69** No. 1–3, 21 (1948)

4  TAMARU, K., 'Pyrolysis and Combustion of Cellulose in the Presence of Inorganic Salts', *Bull. chem. Soc. Japan*, **24** No. 4, 164 (1951)

5  PARKS, W. G., ANTONI, M., PETRARCA, A. E., and PETROCHELLI, A. R., 'The Catalytic Degradation and Oxidation of Cellulose', *Text. Res. J.*, **25**, 789 (1955)

6  SCHWENKER, R. F., and PASCU, E., 'Chemically Modifying Cellulose for Flame Resistance', *Ind. Engng Chem.*, **50** No. 1, 91 (1958)

7  MADORSKY, S. L., HART, V. E., and STRAUS, S., 'Pyrolysis of Cellulose in a Vacuum', *J. Res. natn Bur. Stand.*, **56** No. 6, 343 (1956)

8  MADORSKY, S. L., HART, V. E., and STRAUS, S., 'Thermal Degradation of Cellulosic Materials', *J. Res. natn Bur. Stand.*, **60** No. 4, 343 (1958)

9  HOFMAN, W., OSTROWSKI, T., URBANSKI, T., and WITANOWSKI, M., 'Infrared Absorption Spectra of Products of Carbonisation of Cellulose and Lignin', *Chem. Ind.*, No. 45, 95 (1960)

10  TANG, M. M., and BACON, R., 'Carbonization of Cellulose Fibers: I, Low Temperature Pyrolysis', *Carbon*, **2** No. 3, 211 (1964)

11  MAJOR, W. D., 'The Degradation of Cellulose in Oxygen and Nitrogen at High Temperatures', *TAPPI*, **41** No. 9, 530 (1958)

12  COPPICK, S., *Flameproofing Textiles Fabrics*, ACS Monograph 104 (ed. R. W. Little), Reinhold, New York (1947)

13  GREENWOOD, C. T., KNOX, J. H., and MILNE, E., 'Analysis of the Thermal Decomposition Products of Carbohydrates by Gas-Chromatography', *Chem. Ind.*, No. 46, 1878 (1961)

14  FORD, C. E., and MITCHELL, C. V., and UNION CARBIDE CORP., US Pat. 3 107 152 (15.10.63)

15  SHINDO, A., NAKANISHI, Y., and SOMA, I., 'Highly Crystallite-Oriented Carbon Fibers from Polymeric Fibers', *High Temperature Resistant Fibers from Organic Polymers* (ed. J. Preston), Interscience, New York, 305 (1969)

16  SCHMIDT, D. L., and JONES, W. C., 'Carbon-Base Fiber Reinforced Plastics', *Chem. Engng Prog.*, **58** No. 10, 42 (1962)

17  GREAT LAKES CARBON CORP., Neth. Pat. Applic. 6 408 495 (25.1.66)

18  SHINDO, A., NAKANISHI, Y., and SOMA, I., 'Carbon Fibers from Cellulose Fibers', *High Temperature Resistant Fibers from Organic Polymers* (ed. J. Preston), Interscience, New York, 271 (1969)

19  UNION CARBIDE CORP., US Pat. 3 488 151 (6.1.70)

20  EZEKIEL, H. M., 'Graphite Fibers from an Aromatic Polyamide Yarn', *High Temperature Resistant Fibers from Organic Polymers* (ed. J. Preston), Interscience, New York, 315 (1969)

21  OTANI, S., YOKOYAMA, A., and NUKUI, A., 'Effect of Heat-Treatment under Stress

on MP Carbon', *High Temperature Resistant Fibers from Organic Polymers* (ed. J. Preston), Interscience, New York, 325 (1969)

22  ARAKI, T., and GOMI, S., 'Production of Molten Pitch Carbon Fiber'.*High Temperature Resistant Fibers from Organic Polymers* (ed. J. Preston), Interscience, New York, 331 (1969)

23  OTANI, S., 'On the Carbon Fiber from the Molten Pyrolysis Products', *Carbon,* 3 No. 1, 31 (1965)

24  OTANI, S., 'The Fundamental Structure of MP Carbon Fiber', *Carbon,* 3 No. 2, 213 (1965)

25  OTANI, S., 'On the Raw Materials of MP Carbon Fiber', *Carbon,* 4 No. 3, 425 (1966)

26  OTANI, S., 'Suitable Pitch Material for MP Carbon Fibre', *Carbon,* 6 No. 2, 217 (1968)

27  KAWAMURA, K., and JENKINS, G. M., 'A New Glassy Carbon Fibre', *J. Mater. Sci.,* 5 No. 3, 262 (1970)

28  COAL INDUSTRIES (PATENTS) LTD, Germ. Pat. 1944908 (12.3.70)

29  ADAMS, L. B., BOUCHER, E. A., COOPER, R. N.. and EVERETT, D. H., 'Preparation, Structure and Properties of Saran Carbon Fibres and Powders', Preprint of Paper 7.13, *3rd Conf. Industrial Carbons and Graphite, 1970* (to be published by Society of Chemical Industry)

30  THE CARBORUNDUM CO., Germ. Pat. 1948415 (9.4.70)

31  KOTLENSKY, W. V., TITUS, K. H., and MARTENS, H. E., 'Young's Modulus of Hot-Worked Pyrolytic Graphite', *Nature, Lond.,* 193 No. 4820, 1066 (1962)

32  BACON, R., and SMITH, W. H., 'Tensile Behaviour of Carbonised Rayon Filaments at Elevated Temperatures', *Proc. 2nd Conf. Industrial Carbon and Graphite,* Society of Chemical Industry. London, 203 (1965)

33  BACON, R., and SCHALAMON, W. A., 'Physical Properties of High Modulus Graphite Fiber Made from a Rayon Precursor', *High Temperature Resistant Fibers from Organic Polymers* (ed. J. Preston), Interscience, New York, 285 (1969)

34  RULAND, W., 'The Relationship between Preferred Orientation and Young's Modulus of Carbon Fibers', *High Temperature Resistant Fibers from Organic Polymers* (ed. J. Preston), Interscience, New York, 293 (1969)

35  HUGO, J. A., PHILLIPS, V. A., and ROBERTS, B. W., 'Intimate Structure of High Modulus Carbon Fibres', *Nature, Lond.,* 226 No. 5241, 144 (1970)

36  EZEKIEL, H. M., and SPAIN, R. G., 'Preparation of Graphite Fibers from Polymeric Fibers', *J. Polym. Sci., C,* No. 19, 249 (1967)

37  SCHULMAN, S., 'Methods of Single Fiber Evaluation', *J. Polym. Sci., C,* No. 19, 211 (1967)

38  GIBSON, D. W., and LANGLOIS, G. B., 'Method for Producing High Modulus Carbon Yarn', *Polymer Preprints,* 9 No. 2, 1376 (1968)

39  US AIR FORCE, US Pat. 3399252 (27.8.68)

# 4

# Production of high-modulus fibres from polyacrylonitrile precursor

## 4.1 INTRODUCTION

The work described in the previous chapter indicates that it is possible to produce from cellulose or rayon a graphite fibre having a tensile modulus of at least $50 \times 10^6$ lbf/in². It shows that the modulus is closely related to the degree of orientation of the graphite layers in the fibre structure and that, for a high modulus, these must lie substantially parallel to the axis of the fibre. Whereas partial molecular orientation exists in a rayon precursor, in which some orientation of the graphitic layers is subsequently produced when it is heat treated to 2700°C or over, the alignment is nevertheless insufficient to give the desired tensile modulus. To achieve very high values, hot stretching has to be carried out during the graphitising stage of processing. Whilst this technique is capable of giving the required properties in the finished product, it is nevertheless a difficult one to apply to a production process. The majority of other commercially available precursor materials do not appear to be better than rayon in this respect, with the exception of polyacrylonitrile.

Polyacrylonitrile, in filament form, can be produced with a high degree of molecular orientation. Moreover, this material may be heated to decomposition temperatures without melting to give greater yields of carbon fibre than is the case with rayon. It was therefore considered promising as a precursor for the production of high-modulus carbon fibre. Work carried out to date has indicated that polyacrylonitrile (PAN) is indeed one of the most promising substances for the manufacture of carbon fibres for reinforcement.

In the first part of this chapter, the important historical developments relating to PAN-based carbon fibres are given, including some of the structural investigations which have been carried out. The next part considers the structural changes which take place when PAN fibre is decomposed by heat, together with the requirements necessary to obtain high strength and modulus in the resulting fibre. Finally, a general description is given of the production of PAN-based carbon fibres.

## 4.2    HISTORICAL BACKGROUND

Some of the earliest uses of PAN as a precursor are reported by Shindo, Fujii, and Sengoku, of the Japanese Bureau of Industrial Technics[1], who heated fibres, fabrics, etc., made from a copolymer containing at least 30% acrylonitrile, to 350°C in an oxidising atmosphere, and then to 800°C or over to give a carbon or graphite product of similar shape to the starting material.

This work commenced around 1959. Since then much further work has been reported in the literature, but it is during the last five years that the most notable progress has been made in the production of truly high-strength high-stiffness carbon fibre based on PAN. The most important developments are outlined below.

### 4.2.1    WORK OF SHINDO

Akio Shindo[2] showed that carbon filaments having a tensile modulus up to $24 \times 10^6$ lbf/in$^2$ could be produced from a precursor. Moreover, he showed that the yields of carbon fibre were about 50% by weight, which was much higher than the 10–15% usually obtained from cellulose. It is interesting to note that the relatively high modulus was obtained without any stretching of the fibre during the heat-treatment stages, and Shindo reported that the fibres possessed an appreciable degree of preferred orientation.

The method of conversion involved a pre-oxidation stage. Shindo showed that, whereas PAN began to lose weight at 280°C if heated in a nitrogen atmosphere, pre-oxidation reduced the rate of weight loss. He showed that rapid heating of PAN resulted in the scission of the main chains in the polymer; more gentle heating raised the decomposition temperature and reduced the total loss of weight to give fibres having good flexibility. Thus, slow heating to 1000°C gave an ultimate tensile strength of 106000 lbf/in² for carbon fibre produced from a homopolymer, and 141000 lbf/in² for carbon fibre produced from a copolymer. In these experiments, the original precursor had strengths of approximately 44000 lbf/in², the material being in the form of filaments of 1·5–2·0 denier.

Shindo examined PAN precursor fibres heated to 1000°C by x-ray diffraction, electrical conductivity measurements, and chemical analysis. He showed that, with the pre-oxidised fibre, the hydrogen and nitrogen content of the PAN decreased gradually at first and rapidly at temperatures above 400°C. Hydrogen cyanide was evolved in increasing quantities as the temperature was raised, but the evolution of ammonia was large at the lower temperature only, i.e. up to 500°C. The x-ray diffraction measurements showed that PAN heated to 200°C was virtually unchanged in terms of structure; at 300°C the presence of carbon could be detected, while the measurements in the range 400°–1000°C gave patterns which indicated the presence of polyaromatic groups with ring planes oriented parallel to the fibre axis. Shindo reported that the pyrolysis occurred in two stages: in stage one, pyridine rings were formed at up to 350°C; stage two involved carbonisation at higher temperatures. An important feature of the pre-oxidation stage was that it allowed the carbonisation to commence at lower temperatures. He found that the ultimate tensile strength of pre-oxidised PAN fibre decreased from about 43000 lbf/in² to a minimum value of approximately 14000 lbf/in² at 300°C. The strength then began to increase rapidly. PAN without pre-oxidation showed the same type of behaviour, but the minimum strength occurred at 350°C. The tensile modulus of pre-oxidised fibre remained constant up to 300°C and then increased rapidly, indicating the formation of a carbon structure with good orientation, which was confirmed by electron microscopy and electron and x-ray diffraction measurements.

### 4.2.2    INVESTIGATIONS OF WATT, PHILLIPS, AND JOHNSON, AT THE ROYAL AIRCRAFT ESTABLISHMENT, FARNBOROUGH

The work of Watt, Phillips, and Johnson was aimed specifically at the production of carbon fibres having very high strength and stiffness. They heated PAN fibres in an oxidising atmosphere in the temperature range 200°–250°C and then in a non-oxidising atmosphere up to 1000°C or above. Heating to 1000°C was carried out at the slow rate of 15°C/h in a hydrogen atmosphere to give a carbonised product having a density of 1·6 g/cm³, an ultimate tensile strength of 135000 lbf/in², a tensile modulus of 13 × 10⁶ lbf/in², and hence a breaking strain of $1.0\%$. Additional experiments involved heating PAN fibres for 24 h in air at 220°C to give a tensile strength of 99 000 lbf/in$^2$, followed by heating to 1000°C in a non-oxidising atmosphere and finally to 2500°C to graphitise the filaments, after which the filaments had a tensile strength of 248 000 lbf/in$^2$.

The above investigations carried out at Farnborough form the basis of a National Research Development Corp. (NRDC) patent[3] for the production of high-modulus carbon fibres from PAN precursor. As in the case of Shindo, the essential feature of the work by Watt and his colleagues was an oxidation stage in which the fibre was held under tension, thereby modifying the PAN structure and allowing the subsequent carbonising process to give an improved product. Additional details of the Farnborough work are given in a paper by Watt, Phillips, and Johnson[4], together with properties of their PAN-based carbon fibre. The work carried out by Shindo and at the Royal Aircraft Establishment (RAE), Farnborough, thus showed that the orientation of the graphite crystals in the carbon fibre was the all-important factor, as in the rayon-based material. In fact, their findings confirmed that the mechanical properties of a carbon fibre are largely determined by its crystal structure, irrespective of the starting material. However, in the case of PAN-based fibres, the use of an oxidation stage with tension applied to the fibre prior to carbonising was shown to make a major contribution to the establishment of good crystal orientation in the final product.

### 4.2.3  ADDITIONAL FINDINGS OF WATT AND JOHNSON

Watt and Johnson[5] report the effect of length changes during the oxidation of PAN fibres on the tensile modulus of the resulting carbon fibres. They showed that different tensions applied during the oxidation of the PAN fibre produced a shrinkage or extension during the oxidation stage, and the resulting length changes had a considerable bearing on the final modulus of carbon fibre made from the oxidised material. Examination of the carbon fibre structure by electron microscopy confirmed that an increase of the tensile modulus was accompanied by a greater degree of preferred orientation of the crystal structure. Finally, Watt and Johnson discussed the probable structural changes which take place during the decomposition of PAN fibre from which they derived the following conclusions:

1. A number of crosslinking reactions between molecular chains is possible in PAN fibres heated to 220°C in vacuum or air.

The different reactions proceed simultaneously and depend on the spatial arrangement of the reacting groups.

2. The crosslinking reactions give rise to six-membered rings from which arise graphite nuclei with the basal planes parallel to the direction of the polymer chains.

3. The modulus of carbon fibres made from oxidised PAN fibres increases with the increase in length of the fibres during oxidation. That is because, the more parallel the molecular chains are to the fibre axis during crosslinking reactions, the more oriented will be the graphitic nuclei.

4. The graphite crystallites in the heat-treated carbon fibres are small and turbostratic with an $L_c$ dimension determined by the height of the primary fibrils since crosslinking across fibrils is unlikely on spatial considerations. The crystallites are arranged in what may be called graphitic fibrils derived from the primary fibrils of the parent fibre.

### 4.2.4  ACTIVITIES OF ROLLS-ROYCE LTD

Rolls-Royce Ltd have been interested in the use of strong stiff low-weight fibre reinforcement for many years, one primary area of application being aero-engine gas turbine blades. Quite independently of the RAE, Rolls-Royce carried out work on the production of PAN-based carbon fibres, some of which is reported below.

Standage and Prescott of Rolls-Royce[6] heated PAN fibres to 1 000°C at a rate of less than 1°C/min to produce a graphitic high-modulus form, the heating being carried out in an inert atmosphere. The authors report that filaments of Courtelle (PAN) were heated in nitrogen to 1000°C in 24 h and then to 2700°C at a rate of rise of 30°C/min. The resulting graphite filaments had an average tensile modulus of $32 \times 10^6$ lbf/in². However, when the filaments were heated to 1000°C at 5°C/min and then to 2700°C at 40°C/min, they had a much lower tensile modulus of $7.6 \times 10^6$ lbf/in². The authors also carried out experiments with other sources of PAN fibre, including Acrilan, Zefran, and Orlon, and obtained relatively high values of tensile modulus. The initial heating rates, during carbonising, of 1°C/min therefore proved much more effective than those of 5°C/min, and, in general, fast heating is almost certain to cause deterioration of the structure of the fibre.

Johnson[7] has shown that the strength of carbon fibres made from PAN precursor is governed to a considerable extent by the presence of small flaws both within the fibre and on its surface. Many of the internal flaws originate from the precursor, but other flaws may be created during the heat-treatment stages. Johnson states that incorrect processing conditions such as fast heating rates can maximise the occurrence of voids. Rolls-Royce have attached considerable importance to flaws and structural defects in the precursor and finished fibre since there is little doubt that ultimate properties depend largely on the perfection of the structure. Thus, Johnson and Thorne[8] have devised a method of detecting flaws by a specialised technique of fibre etching. The effect of flaws is discussed in Section 5.4 of this monograph.

Rolls-Royce[9] describe a process for the production of carbon fibres from PAN precursor, using a furnace for oxidation under controlled tension at temperatures up to 265°C, followed by a second furnace with an inert atmosphere and capable of reaching 2500°C. Rolls-Royce have also carried out a considerable amount of work on developing production techniques for making carbon fibre in quantity, and this work has led them to take out a number of patents. One patent[10] describes the production on a continuous basis of carbon-fibre sheets from PAN precursor. After oxidation, the sheets are carbonised continuously in a furnace capable of reaching 1600°C, and this is followed by continuous graphitising in a second furnace if the high-modulus form is required. Rolls-Royce patents[11, 12] also describe continuous processing techniques used in the manufacture of carbon fibres and carbon-fibre sheet and give some details of temperature conditions for processing. A further development by Rolls-Royce[13] has been concerned with the production of high-strength and high-modulus carbon fibres in which hot stretching is carried out at the graphitising stage in order to give further alignment of the graphitic planes to attain an even higher tensile modulus.

### 4.2.5  OTHER WORK ON PAN-BASED CARBON FIBRES

A number of other companies and workers has also been active in the field of work aimed at a better understanding of the process and the mode of structural change occurring during the transformation from PAN to carbon fibre. Some of this work is reported below, but it must be emphasised that considerable further activity is taking place which will increase the knowledge of the process and pave the way for further improvements in fibre properties.

For instance, the Japanese company Tokai Denkyoku Seizo K.K.[14] produced flexible carbon fibres from an acrylonitrile in the form of a homopolymer or copolymer, prepared from not less than 85% (molar) of acrylonitrile and about 15% of a monovinyl compound. The fibre was prepared for carbonising by heating at a temperature of from 180°C to 550°C in an oxygen-containing atmosphere for a time sufficient to blacken the material. The material was then carbonised at a temperature between 700°C and 1200°C for not less than one hour in an inert atmosphere in order to form the carbon fibre. This development has indicated that the structure of the PAN starting material may have appreciable effect on the subsequent heat-treatment stages which must be used to give optimum fibre properties. Thus, the heat-treatment schedule must be compatible with the characteristics of the starting material.

Whereas air oxidation is by far the most widely used technique for combining oxygen with the polymer, it is possible to use instead chemical reagents, such as nitrobenzene. The oxidising reagent may be present in liquid form or may be used as a solid in a fluidised bed in order to effect oxidation. Both Rolls-Royce[15] and Union Carbide[16] have worked in this area, and Union Carbide have described a process involving the oxygenated compounds of one of the transition metals to facilitate reaction. Fibre so oxidised may be subsequently heat treated in the normal manner.

Courtaulds[17] have described in one of their patents the production of carbon fibres from PAN containing cellulose, polyamide, or polyester and involving the usual oxidation, carbonisation, and graphitisation stages, indicating that modifications may be made to the starting material in order to alter the characteristics of the final product.

Fialkov et al.[18] describe structural changes in the heat treatment of PAN fibres, involving carbonisation at 850°–1050°C and graphitisation at 3000°C in an argon atmosphere. This was followed by a steam-activation of one hour. Structure changes were studied by electron microscopy and by the low-temperature nitrogen adsorption–desorption method. Structural anisotropy was observed in carbonised PAN with the presence of an oriented 'fibrillar' structure, having 60–80 Å diameter cylindrical channels between the fibrils. Fibre activation in steam caused marked structural changes, namely the formation of a well-developed channel system owing to the pyrolysis of the interfibrillar reactive non-oriented phase. The specific surface of PAN increased 100-fold owing to activation and the presence of 60 Å diameter channels with 10 Å micropores. The latter disappeared after graphitisation, and a good degree of crystallisation took place.

Work carried out by Brydges et al.[19] of Morganite Research & Development Ltd, on the structure and elastic properties of high-modulus carbon fibres has shown, on the basis of x-ray and electron microscope studies, that such fibres produced from PAN are composed of turbostatic graphite crystallites having a c-axis interlayer spacing of around 3·39 Å, and $L_c$ and $L_a$ measurements of 100 Å and 200 Å respectively. There is a considerable degree of preferred orientation of the crystallites, their c-axes having a mean deviation of about 5° from the fibre axis. Brydges and his co-workers showed that the crystallites appear to be arranged in the form of chains inside fibrils whose diameter is approximately 500 Å. They calculated the upper and lower limits of the elastic moduli of fibres on the assumption that the polycrystals in the fibre were fully dense and contained perfect graphite crystallites which had a measurable amount of preferred orientation. The observed values of the tensile and shear moduli were within the limits corresponding to the assumption of homogeneous stress (Reuss condition, lower limit) or homogeneous strain (Voigt condition, upper limit) among the crystallites. Their result suggested that the tensile modulus of about $150 \times 10^6$ lbf/in$^2$ which is calculated for a fully oriented fibre will also be approached by fibres with preferred orientation only slightly more complete than that of presently available fibres.

Logsdail[20] describes the activities of the Atomic Energy Research Establishment (AERE), Harwell, on the production of carbon fibres based on PAN precursor and using essentially the same process as that developed by Watt, Phillips, and Johnson at the RAE, Farnborough. A Type I high-modulus fibre was produced at the AERE where the heat treatment was extended to 2700°C giving a tensile modulus of $60 \times 10^6$ lbf/in$^2$ and an ultimate tensile strength of 300 000 lbf/in$^2$. In addition, a Type II high-strength fibre was produced for which the heat-treatment temperature was 1500°C, resulting in a modulus of $40 \times 10^6$ lbf/in$^2$ and a strength of 450 000 lbf/in$^2$. Logsdail also describes methods used at the AERE for testing carbon fibres, including non-destructive techniques.

In Japan, Miyamichi et al.[21] carried out extensive studies of the structural changes which occur during the heating of PAN, particularly through the carbonising temperature range. Their work once again confirmed the complexity of chemical changes occurring during the decomposition of PAN and showed that variation of heating rates and furnace atmospheres has a profound effect on the breakdown of the organic molecules and the formation of a fine carbon structure.

## 4.2.6   SUMMARY OF REPORTED INVESTIGATIONS

From the investigations reported in Sections 4.2.1–4.2.5, one of the facts to note is the activities of Shindo, who introduced an oxidation stage prior to carbonising; this was found to improve the quality of the final product in terms of mechanical properties. The most significant work, however, was that carried out by Watt, Phillips, and Johnson at the RAE, Farnborough, who introduced the concept of holding the fibre under controlled tension during the oxidation stage. As a result of their work, it became possible to maintain alignment of the original polymer molecules prior to carbonisation, and this allowed the attainment of truly high-strength and high-modulus fibre without the need for stretching during the final graphitisation stage. The work of Rolls-Royce highlighted a number of important aspects associated with the heat-treatment stages. In particular, Rolls-Royce were concerned with the importance of establishing the best possible fibre structure and they made extensive studies of imperfections. Their work will undoubtedly help to pave the way for further advances.

## 4.3   CHANGES TAKING PLACE DURING THE DECOMPOSITION OF PAN FIBRE

As already stated, changes which take place during the decomposition of PAN fibre are complex and much further work will be necessary to arrive at a complete picture of the carbon fibre structure and the effect of flaws, defects, etc. Equally, the importance of structure perfection in the precursor must be borne in mind in the search for further improvement in carbon fibre properties.

### 4.3.1   STRUCTURE OF PAN FIBRE

Based on the wide range of work already indicated above, a brief description is given of the structure of PAN and the important changes which take place when this material in fibre form is subjected to heat treatment. PAN consists of long chains, as shown in Figure 4.1. It is well known that the chain is fully repetitive and contains CN groupings which have a triple valency bond and are therefore very active. When PAN is heated, a ladder polymer consisting of six-membered rings is formed owing to the linkage between the groups, as shown in the figure. The ladder polymer is more stable towards heat than the original chain and therefore does not melt easily, which explains why PAN is suitable for conversion into carbon fibre by heat treatment. It is interesting to note that PAN fibres heated above 80°C pass through a transition stage and the molecular chains start to shrink. The chains also become distorted and bend and twist in the fibre structure, and orientation is lost, as shown in Figure 4.2. If tension is now applied to the fibre during the heating, the molecular chains are prevented from shrinking and their orientation parallel to the fibre axis is therefore preserved. If the tension is removed from the fibre and the latter is reheated, shrinkage will occur once more giving rise to lack of orientation. It should be noted that the chains themselves tend to be bonded together to form aggregates as a result of the active CN

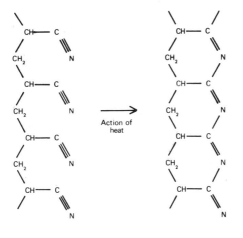

(a) PAN chain                    (b) Ladder polymer

*Figure 4.1. Molecular structure of PAN chains: (a) the molecular arrangement of chains of PAN; (b) the change which takes place due to the action of heat*

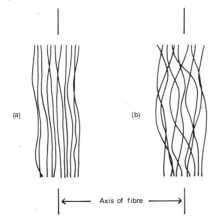

← Axis of fibre →

*Figure 4.2. Effect of heating on PAN fibre molecular chains: (a) PAN fibre molecular chains lying substantially parallel to the fibre axis in unheated filaments; (b) heated PAN fibre showing bent and twisted molecular chains (their orientation to the fibre axis is reduced and the fibre undergoes shrinkage)*

D

groups. These aggregates are usually called 'fibrils'. It is likely that the formation of the fibrils takes place during the precipitation of the PAN in the actual spinning process, and it is believed that they can have a branched structure such that one may be bonded to another via molecular chains common to each.

### 4.3.2    EFFECT OF HEAT ON PAN FIBRE

When PAN is heated in a vacuum or air to over 200°C, a number of chemical reactions commence resulting in the formation of hydrogen cyanide, ammonia, and other by-products which are liberated as volatiles. Some chemical bonding takes place between the molecular chains, probably via the active CN groups, which results in the formation of a stronger chemical bond than the one causing the initial aggregation on spinning. However, these bonds are not particularly strong by normal chemical standards. On the other hand, if PAN is heated in an oxidising atmosphere to about 220°C for a sufficient time, a chemical reaction takes place in which oxygen is taken up into the fibre, as illustrated in a simplified form in Figure 4.3.

The take-up of oxygen is associated with the formation of good chemical crosslinks which firmly bond the molecular chains

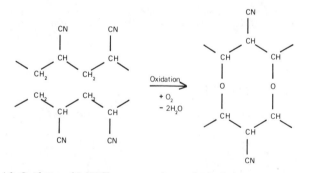

*Figure 4.3. Oxidation of PAN fibre: oxygen forms a bridge linking two PAN molecules and water is liberated*

together. Since the CN groups can be oriented at different angles, several molecular chains can be tied together by the oxygen bonding. This oxygen bond is believed by Watt and Johnson[5] to be capable of spanning the distances between the molecular chains more effectively than CH and CN groups. Theoretical considerations indicate that the maximum amount of oxygen taken up may exceed 15%. In practice, however, the amount absorbed lies in the range 8–10%. Owing to the loss of water during the oxidation stage, the carbon content of the oxidised PAN is increased from around 30% to about 45% and the colour changes from white to black. The chemical changes taking place during the oxidation of PAN are complex. Kesatochkin and Kargin[22] and Grassie, Hay, and McNeil[23] have postulated the significant chemical reactions which occur. For carbon fibre manufacture, however, the important feature is the formation of strong oxygen crosslinks. The effect of these links on the molecular chains of PAN is shown in a very simplified form in Figure 4.4.

The work of Watt and his colleagues[3–5] thus marked a major step forward since they carried out the oxidation of PAN fibre

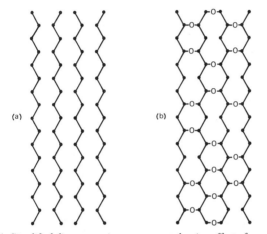

Figure 4.4. Simplified diagrammatic arrangement showing effect of oxygen crosslinks on molecular chains of PAN fibre: (a) molecular chains of PAN fibre lying parallel to each other; (b) oxygen bonds linking PAN chains (the links occur in three dimensions and bond successive adjacent layers of PAN chains so that they cannot bend and distort on heating)

whilst it remained under tension, thereby allowing the formation of the oxygen crosslinks between the molecular chains. Hence, the chains remained straight and oriented parallel to the fibre axis, even when all the tension had been removed on completion of the heating period. Thus, the molecules were positioned in an ideal way for the subsequent conversion into carbon or graphite fibres.

The carbon ring structure produced during carbonisation (see Figure 4.5) has a crystalline form. The crystals are very small but nevertheless show good alignment parallel to the fibre axis owing to their derivation from the oriented polymer chains in the original PAN and the oxidised PAN fibre. Increased amounts of heat

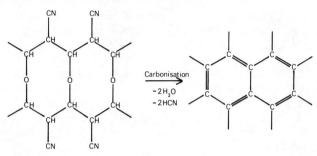

*Figure 4.5. Formation of a carbon ring structure by carbonising oxidised PAN fibre: the carbon rings will have an orientation which is dependent on that of the original PAN chains and their subsequent oxidised form*

treatment, and in particular graphitisation, develop the crystals, reduce fibre porosity, and impart high modulus values. After graphitisation, the crystals are well defined and aligned within a few degrees of the fibre axis.

## 4.4 EFFECT OF HEAT-TREATMENT CONDITIONS ON THE PROPERTIES OF PAN-BASED CARBON FIBRES

It has been shown that carbonised fibre made from oxidised PAN precursor possesses a fine and well-oriented crystal structure which is responsible for its high strength and relatively high modulus. It

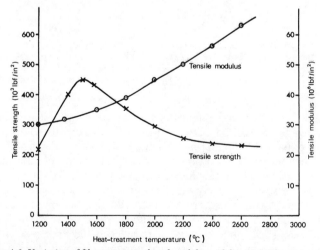

*Figure 4.6. Variation of filament strength and modulus with heat-treatment temperature for PAN-based carbon fibres*

has been stated that graphitisation of carbonised fibre increases the modulus further.

The heat-treatment conditions used have a profound effect on the mechanical properties of PAN-based carbon fibres. Fibres heated to around 1000°C have an ultimate tensile strength in the range 100 000–200 000 lbf/in$^2$ and a tensile modulus below $30 \times 10^6$ lbf/in$^2$. The strength values increase markedly with temperature until a maximum of 400 000 lbf/in$^2$ or over is reached at approximately 1500°C. Above this temperature, the strength gradually falls away until around 2200°C is reached, but heat-treatment temperatures beyond 2200°C do not significantly alter the strength. The tensile modulus of the fibre, on the other hand, shows a steady increase from approximately 1000°C upwards, extending to $60 \times 10^6$ lbf/in$^2$ at 2600°C and above. Figure 4.6 shows the relationship of both strength and modulus to heat-treatment temperature. It will be seen from these graphs that fibre properties may be varied over a relatively wide range by suitable choice of heat-treatment conditions.

## 4.5   MANUFACTURE OF HIGH-MODULUS CARBON FIBRES FROM PAN PRECURSOR

The description given below indicates the general principles involved in the commercial production of these fibres. It must be appreciated, however, that in practice much depends on experience in the operation of the various heat-treatment stages and in the handling of the fibre. Manufacturing companies are not prepared to disclose production techniques for obvious reasons, and hence the outline below is given as a general guide only.

### 4.5.1   RAW MATERIALS

Polyacrylonitrile, usually in the form of a copolymer, is used as a starting material. Almost all the work carried out in the UK has involved Courtelle fibre of circular cross-section manufactured by Courtaulds Ltd. It has been shown that the various heat-treatment stages can be carried out most effectively if the PAN precursor takes the form of very fine filaments; in practice, 1–3 denier filaments are used and these give satisfactory carbon fibre properties. Larger diameter filaments may be used, but the heat-treatment stages will have to be modified accordingly. It is usual to have a relatively large number of individual filaments in a fibre yarn or tow, and most of the British work has been carried out with 10 000 filament tows.

This material has been shown capable of giving particularly good fibre properties, but other sources of PAN may be used successfully.

### 4.5.2   OXIDATION STAGE

As already stated, it is necessary to oxidise the fibre before commencing the carbonising stages. Oxidation takes place at 200°–250°C in air for a period of time sufficient to ensure completion of the reaction which involves the creation of the crosslinks in the organic molecular chains. Application of tension to the fibre during the oxidation stage keeps these molecular chains stretched and oriented parallel to the fibre axis. The degree of applied tension is found by practice, and useful guidelines are given in published work. It should be noted that the time taken for the oxidation reaction is a function of the fibre diameter; large diameters require longer oxidation times. During the oxidation stage, an exothermic chemical reaction takes place and some toxic gases are liberated which include cyanide groups.

Since cyanide groups are highly toxic, the design of an oxidation plant must prevent gas leakage. This may be achieved with suitable extraction ventilation working in conjunction with a scrubber unit to take out toxic gases. For this reason and because the operating temperatures are very low, the oxidation plant is normally constructed in metal. Any form of heat source, for example electricity or gas, may be used, but it is desirable to be able to achieve good temperature control coupled with free circulation of the air to ensure uniform oxidation of the precursor.

Oxidation plants are designed as either batch or continuous units. In the former, the fibre is suitably held under tension during the whole cycle. Continuous plants, on the other hand, are constructed to allow the fibre to be pulled through the oxidation chamber whilst under tension, the speed being calculated to allow adequate oxidation to take place. Since the oxidation tends to be a slow process, efforts have been made to obtain faster conversion rates. Work has been carried out on the manufacture of carbon fibre from doped PAN precursor by Moutaud and Cauville[24], of Le Carbone–Lorraine, who discuss the influence of metal additives to PAN fibres on the subsequent controlled oxidation treatment and make a comparison between doped and non-doped precursor fibres. The essential feature of this investigation is the highlighting of the possible role of metal ions or other additives in catalysing and speeding up the oxidation of the PAN precursor.

### 4.5.3   CARBONISATION STAGE

After oxidation, the fibre is black and has a shiny appearance, but it is still an organic material having a structure resembling that of the original PAN. Further heat treatment progressively converts the material to carbon. During the carbonisation stage, the fibre must be kept out of contact with air, and it is usual to use an inert gas such as nitrogen or argon in the carbonisation furnaces. These furnaces may consist of a heat-resistant metal chamber for temperatures up to say 1000°C, but for higher temperatures some form of refractory material must be used. Heating can be by means of electrical elements or by gas or oil burners. The design of a carbonising furnace is governed largely by the type of plant involved—batch or continuous processing.

This stage of the process involves the change from an organic fibre to carbon, and a range of volatiles is liberated; in fact. up to 50% loss in weight may occur in the fibre, producing large volumes of volatiles and tarry substances. As in the oxidation stage, the volatiles are toxic and care has to be taken to cope with them in much larger quantity. There is a very large weight loss during the carbonisation stage, and the heating rates have to be carefully controlled to ensure that the release of volatiles does not damage the fibre, as the volatiles have to diffuse from the centre of the fibre to the surface. During the carbonisation stage, the fibre shrinks in diameter and in length, most of this shrinkage taking place before 800°C is reached. Most of the volatiles come off before 1000°C, but traces of organic carbon-rich residues remain until 1400°–1500°C. The high-strength fibre results from carbonisation to 1100°–1500°C, the material consisting essentially of very small crystals. These fibre crystals have good orientation, but there is some porosity in the fibres.

Many variants are possible in the design of both oxidation and carbonisation units. For example, Rolls-Royce[25] have modified the heat-treatment process to give optimum properties for gas turbine compressor-blade applications. They indicated an oxidation stage at 150°–300°C followed by carbonisation from 900°–1200°C in an inert atmosphere. The fibres are then incorporated into composites after pre-impregnation with a resin, preferably in epoxy novolak. Use of the lower temperature is said to give improved shear, torsional, and flexural strengths in the composite, although the tensile modulus is somewhat lower than for fibres carbonised at 1500°C. In the production of continuously heat-treated carbon fibre, a number of techniques has been developed to allow the fibre to pass continuously through the heat-treatment stage. Zbrzezniak[26] has developed a technique whereby carbon fibres of high strength and tensile modulus can be produced continuously, the fibre being laid in a cage on rollers placed in the furnace. A modification of the PAN-based carbon fibre manufacturing process is to utilise the basic technique for the production of fabrics from high-strength carbon fibre. Normally, it is difficult to weave carbonised and particularly graphitised material owing to the high modulus and the risk of fibre damage, but, by carrying out the weaving operation after the oxidation stage, it is possible to produce fabrics without any difficulty. The fibres can then be carbonised in the normal manner to 1000°C or even graphitised if required. The NRDC have a patent covering such work[27].

Johnson, Rose, and Scott[28] have shown that flaws and porosity can be produced in carbon fibres heat treated too rapidly. For example, PAN-based carbon fibres subjected to rapid oxidation develop a hard oxidised layer on their outer surface while the interior remains in an unoxidised condition. Further heat treatment carried out on these fibres results in the occurrence of voids and flaws.

### 4.5.4   GRAPHITISATION STAGE

The graphitisation stage is necessary for the production of the high-modulus fibre, and temperatures up to 2800°C are involved. The graphitising is invariably carried out in an electrically heated furnace, using graphite for the high-temperature areas. A major requirement in graphitising furnaces used for carbon fibre manu-facture is the complete exclusion of air since even minute quantities will result in reduced fibre properties. In practice, the choice of furnace atmosphere is very limited and argon gas is the most widely used, but helium may be regarded as a possible alternative. It should be noted that nitrogen cannot be used as it will react rapidly with carbon at temperatures over 2000°C to form cyanogen. A point to note is that heating rates used in graphitising furnaces are often very high, much higher than those used in carbonising, the reasons being that there are virtually no volatiles to remove from the fibre and that the development of the graphite structure in carbon takes place very quickly.

Graphitising furnaces can, once again, be divided into two main types. One type is for batch processing, in which the fibre is loaded and the furnace is taken through a heating cycle to the top tempera-ture and then cooled. The second type is for continuous processing and is designed to remain at working temperature, the fibre being pulled through the high-temperature zone.

## 4.5.5 SURFACE-TREATMENT STAGE

It will be seen in subsequent chapters that the success or failure of carbon-fibre reinforced resins is governed by the degree of adhesion occurring between the fibre surface and the resin matrix. In the case of glass fibre, it is normal practice to 'size coat' the surface of the filaments to provide a better bond, thereby allowing the resin matrix to transmit the stresses from filament to filament in the optimum manner. With carbon fibre, the degree of bonding normally attainable is far less than the strength of the resin, implying that the latter cannot function to the best advantage. This is particularly true for the high-modulus variety since the structure of this material is essentially graphitic, presenting a poor bonding surface to the matrix.

Work has shown that it is possible to surface treat carbon fibre whereby the surface is conditioned to provide improved bonding. In particular, the PAN-based fibres can be treated to give predictable results. A number of methods is available including wet chemical treatments, one of which is reported to use a solution of sodium hypochlorite[29]. Other solutions which have been used are nitric acid, acetic acid, and sulphuric acid. The precise action of the surface treatment is not known, but some degree of chemical cleaning takes place and it is almost certain that a very mild degree of etching is involved together with the creation of active sites on the fibre surface, all of which assist in establishing improved adhesion. In contrast to the sizing of glass fibres, the surface treatment of PAN-based carbon fibre does not involve the application of any film or coating on to the fibre surface. Any process selected has to be carried out with care and demands a strict routine if reproducible results are to be obtained. If the method is carried out correctly, greatly improved adhesion between fibre and matrix is obtained without degrading the properties of the fibre in any way. It should be noted that the techniques which have been developed improve the bond between the fibre surface and thermosetting resins; for certain thermoplastics and possibly metals, it is likely that other processes will have to be developed in order to achieve optimum results.

### REFERENCES

1 SHINDU, A., FUJII, R., and SENGOKU, M. and JAPANESE BUREAU OF INDUSTRIAL TECHNICS, Jap. Pat. 4405/1962 (13.6.62)
2 SHINDO, A., 'Graphite Fibre', *Rep. Osaka ind. Res. Inst.*, No. 317 (1961)
3 NATIONAL RESEARCH DEVELOPMENT CORP., Brit. Pat. 1 110 791 (24.4.68)
4 WATT, W., PHILLIPS, L. N., and JOHNSON, W., 'High-Strength, High Modulus Carbon Fibres', *Engineer, Lond.*, **221** No. 5757, 815 (1966)
5 WATT, W., and JOHNSON, W., 'The Effect of Length Changes during the Oxidation of Polyacrylonitrile Fibers on the Young's Modulus of Carbon Fibres', *High*

*Temperature Resistant Fibers from Organic Polymers* (ed. J. Preston), Interscience, New York, 215 (1969)

6 ROLLS-ROYCE LTD, Brit. Pat. 1 128 043 (25.9.68); this patent corresponds to Belg. Pat. 678 679 (30.3.66) and Fr. Pat. 1 471993 (27.6.67)

7 JOHNSON, J. W., 'Factors Affecting the Tensile Strength of Carbon–Graphite Fibres', *High Temperature Resistant Fibers from Organic Polymers* (ed. J. Preston), Interscience, New York, 228 (1969)

8 JOHNSON, J. W., and THORNE, D. J., 'Effect of Internal Polymer Flaws on Strength of Carbon Fibres Prepared from an Acrylic Precursor', *Carbon,* 7 No. 6, 659 (1969)

9 ROLLS-ROYCE LTD, Fr. Pat. 1 580 443 (5.9.69)

10 ROLLS-ROYCE LTD, Germ. Pat. 1 805 901 (6.11.69)

11 ROLLS ROYCE LTD., Fr. Pat. Applic. 2 006 543 (6.2.70)

12 ROLLS-ROYCE LTD., Fr. Pat. Applic. 2 008 173 (6.3.70)

13 ROLLS-ROYCE LTD, Brit. Pat. 1 174 868 (17.12.69)

14 TOKAI DENKYOKU SEIZO K. K., US Pat. 3 285 696 (15.11.66)

15 ROLLS-ROYCE LTD, Fr. Pat. Applic. 2 003 359 (26.12.69)

16 UNION CARBIDE CORP., Fr. Pat. 1 579 198 (22.8.69)

17 COURTAULDS LTD. Germ. Pat. 1 925 118 (20.11.69)

18 FIALKOV, A. S., BABER, A. J., SIMIRNOV, B. N., and SENENOVA, L. P., 'Structural Changes during Thermal Treatment of Polyacrylonitrile Fibres' (in Russian), *Dokl. Akad. Nauk SSSR,* 173, 147 (1962)

19 BRYDGES, W. T., BADAMI, D. B., JOINER, J. C., and JONES, G. A., 'The Structure and Elastic Properties of Carbon Fibres', *High Temperature Resistant Fibers from Organic Polymers* (ed. J. Preston), Interscience, New York, 255 (1969)

20 LOGSDAIL, D. H., 'Aspects of Carbon Fiber Development at AERE, Harwell', *High Temperature Resistant Fibers from Organic Polymers* (ed. J. Preston), Interscience, New York, 245 (1969)

21 MIYAMICHI, K., et al., 'Carbonisation of Polyacrylonitrile Fibres', *J. Soc. Text. Cellul. Ind. Japan,* 21 No. 12, 640 (1965)

22 KESATOCHKIN, V. I., and KARGIN, V. A., 'Thermal Conversion of Oriented PAN', *Dokl. Akad. Nauk SSSR,* 191, 1084 (1970)

23 GRASSIE, N., HAY, J. N., and MCNEIL, I. C., 'Coloration in Acrylonitrile and Methyacrylonitrile Polymers', *J. Polym. Sci.,* 31, 203 (1958)

24 MOUTAUD, G. M., and CAUVILLE, R. J., 'Carbon Fibre Obtained from Doped PAN', Preprint of Paper 7.12, *3rd Conf. Industrial Carbons and Graphite, 1970* (to be published by Society of Chemical Industry)

25 ROLLS-ROYCE LTD, Germ. Pat. 1 923 622 (12.2.70)

26 ZBRZEZNIAK, J., and MORGANITE RESEARCH AND DEVELOPMENT LTD, Fr. Pat. Applic. 2 012 224 (30.4.70)

27 NATIONAL RESEARCH DEVELOPMENT CORP., Fr. Pat. 1 539 755 (25.9.68)

28 JOHNSON, J. W., ROSE, P. G., and SCOTT, G., Preprint of Paper 7.5, *3rd Conf. Industrial Carbons and Graphite, 1970* (to be published by Society of Chemical Industry)

29 AIKEN, I. D., RHODES, G., and SPENCER, R. A. P., *Development of a Wet Oxidation Process for the Surface Treatment of Carbon Fibres,* UKAEA Research Group, Harwell (1970)

# 5

# Testing and properties of high-modulus PAN-based carbon fibres

## 5.1 INTRODUCTION

The first part of this chapter is concerned with the description of proven test procedures for the measurement of the mechanical properties of carbon fibres. Methods of testing are important as they play a decisive role in determining final results obtained; moreover, work carried out in one location can only be compared effectively with that carried out elsewhere if test procedures can be repeated and reproduced at will.

Based on established test methods, the mechanical properties of PAN-based carbon fibres are then outlined and comparison is made with other fibres and reinforcement materials. In the last part of the chapter, consideration is given to the structure of PAN fibre in relation to its properties and the opportunity is taken to discuss some recent research work aimed at identifying more clearly the role of the fibre structure and at securing even better mechanical properties.

## 5.2 TESTING OF CARBON FIBRES

The testing of carbon fibres is highly specialised and requires very careful handling techniques if reliable results are to be obtained.

The support and mounting of specimens in a satisfactory and consistent manner is one of the major difficulties arising from the very small diameter of the fibres. The methods described below are based on the experience at Morganite Modmor Ltd and have been presented in a paper by Blakelock and Lovell[1]. The general principles involved have found widespread acceptance for the testing of carbon fibres.

### 5.2.1 MEASUREMENT OF FIBRE MODULUS AND ULTIMATE TENSILE STRENGTH

The most direct method of measuring tensile modulus and ultimate tensile strength is on individual filaments, in which case it is essential to carry out measurements on a number of filaments and average the results to give a mean value. The reason for this is that any filament may have some minor flaw or defect which will lower the strength but the testing of a number of filaments averages out the effects and enables a more representative value to be obtained.

Single filaments are selected for mounting on to a card, by carefully pulling them out of the sample tow which is previously cut to approximately 6 in in length. A piece of adhesive tape is used to remove the single filament, which would be too small to be moved by hand, as shown in Figure 5.1(a). The single filament is then pulled across the card, as illustrated in Figures 5.1(b)–5.1(d), an adhesive such as Durafix being used to hold the filament. The adhesive must be allowed to harden before the filament is used, otherwise slipping may occur.

It will be seen that the card has a large rectangular and two small circular holes and that the filament is stretched across all three. The circular holes are for the purpose of fibre diameter measurements, while the rectangular hole is intended for the portion of the fibre which will break.

An Instron tester is ideal for this kind of work, but it requires the fitting of a calibrated load cell to measure the load applied to the fibre. In practice, a full electrical output from the load cell corresponding to full-scale deflections of the associated strip chart recorder should occur at 10–20 g load.

The rectangular portion of the card is mounted in the Instron machine as shown in Figures 5.2(a) and 5.2(b). The grips used are free-floating and exert tension along the fibre axis. This is of great importance if reliable results are to be obtained as non-axial loading causes bending stresses and lower strength values. For this reason it is also essential that the card be mounted centrally in the grips. Figures 5.2(c) and 5.2(d) respectively show the pieces of the card being cut off and the fibre specimen finally mounted between the jaws of the tester. At this stage the load can be applied after any slack has been taken up.

Application of the load produces a straight line on the recorder, as shown in Figure 5.3. The horizontal distance *l* across the chart is an indication of the applied load at the point of fracture of the

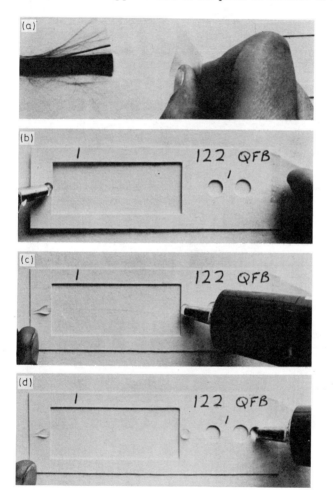

*Figure 5.1. Mounting a single filament for testing: (a) use of adhesive tape to remove a single filament from the fibre tow; (b) first stage of stretching and sticking a single filament to the card using Durafix adhesive; (c) second stage of sticking a single filament to the card; (d) final stage of sticking a single filament to the card*

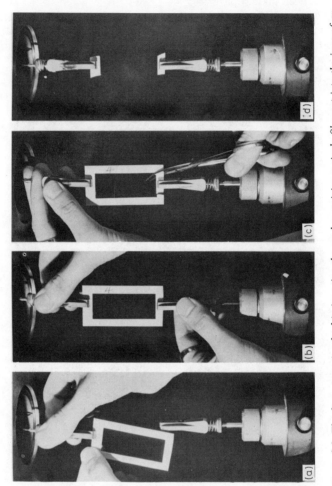

Figure 5.2. The four stages involved in inserting the card containing a single filament into the jaws of an Instron tester[1]: the card is mounted centrally in the jaws of the machine, after which the card on either side of the jaw is cut away so that only the filament remains suspended between the two jaws

specimen. During application of the load, the filament stretches; it does so as a perfectly elastic body, unlike metals, as shown by the straight-line trace on the chart. The cotangent of angle $\alpha$ is a measure of the load–extension ratio.

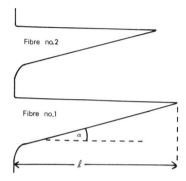

*Figure 5.3. A typical strip chart recording obtained in an Instron tester, showing a straight-line trace once the load on the filament has been taken up; the other line indicates the fly-back after the specimen has broken; the length l and the cotangent of the angle $\alpha$ indicate respectively the breaking load and the load–extension ratio*

In order to determine the filament modulus and strength, its diameter must be known with high accuracy. Two well-known methods can be used for measuring it: one is an optical method which employs the Watson Image Shearing Eyepiece fitted to a normal microscope; the other involves measuring the weight of the fibre per unit length and its density. Where carbon filaments are being tested as part of process control on manufacture, one may assume that fibre density remains constant and hence the second method is able to provide a quick answer. Density may be checked at regular intervals by a floatation method using a carbon tetrachloride–bromoform mixture. An advantage of the optical method is that local variations of fibre diameter may be observed if they occur. An important limitation of optical methods arises, however, when the fibre does not have a circular cross-section, and in such an event one may have to apply correction factors or estimate the diameter. In the case of British PAN-based carbon fibre using Courtelle as the precursor, the end-product has a circular cross-section, but other PAN sources may give different sections, whilst rayon gives an irregular cross-section.

It is the practice of Morganite Modmor to use a gauge length of 5·0 cm for the Instron tester. Gauge length has an important bearing on the ultimate tensile strength values obtained—the greater the length, the lower the ultimate tensile strength, owing to the fact that imperfections in the fibre are more likely to occur in a long length than in a short length. Figure 5.4 shows the effect of gauge length on fibre strength. It will be seen that the curve for the high-modulus fibre tends to flatten out much more than for the high-strength form. It is evident that the filament strengths are still falling with a gauge length of 20 cm, which in turn suggests that a range of

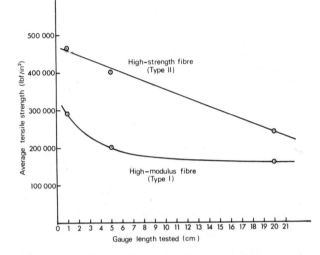

*Figure 5.4. Variation of average strength of 50 fibres with testing gauge length*

possible defects exists in PAN-based carbon fibre, the full complement of which is normally spread over at least such a fibre length. A similar situation exists with rayon-based carbon fibres. The reasons for this behaviour are not fully understood, but it appears that defects occur in the fibre structure or at the fibre surface, becoming more numerous as the length of the specimen is increased and therefore accounting for a lower tensile strength.

## 5.2.2   MEASUREMENT OF FIBRE PROPERTIES BY OTHER METHODS

In the continuous manufacture of carbon fibre, it is useful to be able to measure mechanical properties continuously since this will give an immediate indication of any change in the process. Rolls-Royce Ltd[2] have developed a method which depends on the fact that, as the tensile modulus of fibre increases, the resistivity decreases. Hence, from measurement of the electrical resistivity of the fibre, a good indication of its modulus value can be obtained. In the Rolls-Royce equipment, fibre is stretched between rollers and passed through a coil connected to a tuned oscillator, which allows direct changes in resistivity to be measured as the fibre passes through the coil. Reynolds[3] at the Non-Destructive Testing Centre, AERE, has also looked into a method of measuring modulus on a continuous basis by determining the transit time for short ultrasonic pulses along single fibres. It is believed that the technique may be capable of further development to allow 10 000 filament tow to be continuously monitored for its modulus value. The AERE are also developing semi-continuous measurement of modulus by an x-ray diffraction method.

## 5.3    PROPERTIES OF PAN-BASED CARBON FIBRE

### 5.3.1    PROPERTIES OF MODMOR FIBRE

*Table 5.1* gives the properties of PAN-based Modmor fibre. The values given were obtained by the test methods described in this chapter and may be taken as an indication of the performance of

**Table 5.1** TYPICAL MODMOR FIBRE PROPERTIES

| Property | High-modulus fibre (Type I) | High-strength fibre (Type II) |
|---|---|---|
| Ultimate tensile strength (lbf/in$^2$) | 250 000 | 350 000 |
| Tensile modulus ($10^6$ lbf/in$^2$) | 60 | 35 |
| Diameter ($10^{-3}$ in) | 0·3 (8 $\mu$m) | 0·3 (8 $\mu$m) |
| Density (lb/in$^3$) | 0·072 | 0·063 |
| Interlaminar shear strength (lbf/in$^2$) | 8000 | 11 000 |

PAN-based materials. In particular, the properties of PAN-based carbon fibres produced by the RAE, Farnborough, and by Courtaulds Ltd (trade mark, Grafil) and published in technical literature are similar.

### 5.3.2    VARIATION OF PROPERTIES OF PAN-BASED FIBRES AND COMPARISON WITH OTHER REINFORCEMENT MATERIALS

*Table 5.2* shows the coefficient of variation for values of fibre strength, modulus, and diameter, based on a large number of tests carried out on Modmor fibres. From the data presented in *Table 5.2*, which refers to the testing of 40 filaments in a batch, the mean accuracy of measurement achieved for 95% confidence level can be derived. This level of confidence is also indicated in the range of

**Table 5.2** COEFFICIENT OF VARIATIONS AND LEVELS OF ACCURACY FOR BOTH SHORT AND CONTINUOUS MODMOR FIBRE

| Property | Short-length fibre (1m long) | | Continuous fibre | |
| | Coefficient of variation within a batch of specimens; range of values from individual specimens within a batch is given in brackets (%) | Mean accuracy of measurement; (95%, confidence values) for 40 fibres tested)* (%) | Coefficient of variation within specimens; range of values from individual specimens in brackets (%) | Mean accuracy of measurement; (95% confidence values for 40 fibres tested )* (%) |
|---|---|---|---|---|
| **Strength** | 28 (26–30) | 8·7 | 22 (14–30) | 6·8 |
| **Modulus** | 10 (9–11) | 3·1 | 10 (4–16) | 3·1 |
| **Diameter** | 7 (6–8) | 2·2 | 8 (6–10) | 2·5 |

**Table 5.3** CONSISTENCY OF PROPERTIES OF MODMOR CARBON FIBRES

| | High-modulus fibre (Type I)* | | | | High-strength fibre (Type II)† | | | |
|---|---|---|---|---|---|---|---|---|
| | Ultimate tensile strength (lbf/in²) | Tensile modulus (10⁶ lbf/in²) | Diameter (μm) | Interlaminar shear strength (lbf/in²) | Ultimate tensile strength (lbf/in²) | Tensile modulus (10⁶ lbf/in²) | Diameter (μm) | Interlaminar shear strength (lbf/in²) |
| Mean | 246 000 | 60·78 | 7·8 | 8900 | 401 000 | 37·6 | 8·1 | 11 800 |
| Minimum | 196 000 | 43·9 | 7·0 | 7100 | 344 000 | 34·0 | 7·6 | 8 500 |
| Maximum | 296 000 | 68·3 | 9·7 | 10 500 | 462 000 | 40·7 | 8·6 | 14 400 |
| Standard deviation | 23 000 | 4·1 | 0·33 | 1030 | 30 800 | 1·8 | 0·2 | 900 |
| Coefficient of variation | 9·5% | 6·7% | 4·3% | — | 7·7% | 4·9% | 2·9% | — |

\* Density, 0·07 lb/in³; breaking strain, 0·41%; gauge length tested, 5 cm.
† Density, 0·063 lb/in³; breaking strain, 1·07%; gauge length tested, 5 cm.

**Table 5.4** TYPICAL PROPERTIES OF HIGH-MODULUS CARBON FIBRES COMPARED WITH OTHER REINFORCING MATERIALS

| Reinforcement | Ultimate tensile strength ($lbf/in^2$) | Tensile modulus ($10^6\ lbf/in^2$) | Density ($lb/in^3$) | Specific ultimate tensile strength ($10^6$ in) | Specific modulus ($10^6$ in) |
|---|---|---|---|---|---|
| *Fibres* | | | | | |
| Rayon-based carbon (Thornel 25) | 175 000–200 000 | 24–28 | 0·053 | 3·5 | 490 |
| Rayon-based carbon (Thornel 50) | 275 000–300 000 | 44–55 | 0·060 | 4·8 | 830 |
| PAN-based carbon (Type I) | 225 000–275 000 | 55–60 | 0·072 | 3·5 | 800 |
| PAN-based carbon (Type II) | 325 000–375 000 | 32–38 | 0·063 | 5·6 | 560 |
| E-glass fibre | 400 000–500 000 | 10–11 | 0·092 | 4·9 | 115 |
| S-glass fibre | 500 000–700 000 | 12–13 | 0·092 | 6·5 | 135 |
| Silica glass | 800 000–850 000 | 10–11 | 0·092 | 9·0 | 115 |
| *Composite fibres* | | | | | |
| Boron on tungsten base | 400 000–500 000 | 55–60 | 0·095 | 4·7 | 600 |
| *Wires* | | | | | |
| Beryllium | 130 000–140 000 | 35–40 | 0·066 | 2·0 | 560 |
| Tungsten | 500 000–600 000 | 48–52 | 0·7 | 0·8 | 70 |

the quoted mean, plus and minus the figures given in the columns in *Table 5.2* marked with an asterisk. The testing of a few filaments per batch, say five or 10, greatly reduces the level of confidence obtained. *Table 5.3* gives further data relating to the consistency of properties of Modmor carbon fibres, while *Table 5.4* compares the typical properties of a number of high-modulus carbon fibres with other reinforcing materials.

### 5.3.3   DISTRIBUTION OF PROPERTIES WITHIN A BATCH

Figures 5.5 and 5.6 show respectively the distribution of strengths and moduli within a batch of Type I Modmor fibre*. It will be seen that the spread of strength is greater than that of modulus. This is invariably the case and easily explained by the fact stated earlier that defects affect the strength but not the modulus. If the testing is carried out over a shorter gauge length, the actual ultimate tensile strength values will be increased as shown in Figure 5.4, and moreover the spread of results will tend to decrease. However, gauge length does not affect tensile modulus values.

### 5.3.4   EFFECT OF TEMPERATURE ON THE PROPERTIES OF CARBON FIBRES

It is significant that carbon or graphite fibres possess exceptionally good high-temperature properties provided they are kept out of contact with either air or other oxidising atmospheres. Figure 5.7 shows the general relationship between strength and temperature for carbon fibres, whether based on PAN or rayon, and for other structural and reinforcing materials. The curves show quite clearly that all the other materials exhibit a sharp fall-off in strength as temperature is increased and that above 1500°C carbon-based fibres are supreme. This characteristic may have relatively little significance where low-temperature applications are envisaged, but, in combination with high-temperature matrices, carbon fibres appear to show more promise than any other known materials.

### 5.3.5   STRESS–STRAIN CHARACTERISTICS OF CARBON FIBRE

An important feature of carbon fibre materials of the truly high-modulus or high-strength type is that they are elastic all the way to failure, in the same way as glass fibres. However, the degree of elongation of the filaments at maximum (breaking) load is very

---

* Type I fibre refers to material which has been graphitised to yield maximum stiffness.

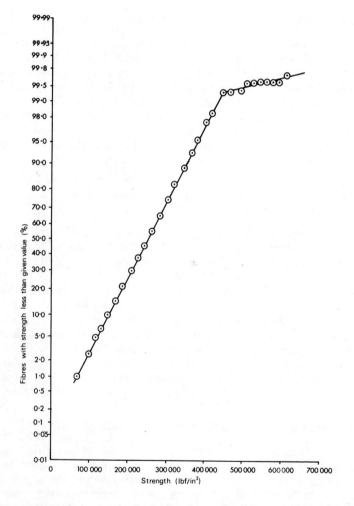

*Figure 5.5. Graph showing the distribution of strengths of filaments within a batch of Type 1 Modmor fibre*

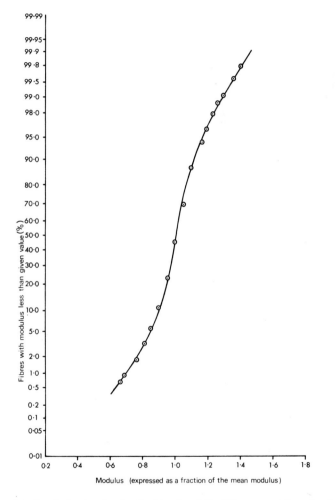

*Figure 5.6. Graph showing the distribution of moduli of filaments within a batch of Type 1 Modmor fibre*

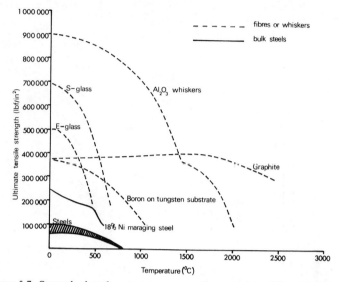

Figure 5.7. Strength plotted against temperature for a number of fibre or whisker reinforcing materials compared with bulk steels and a high-temperature steel alloy

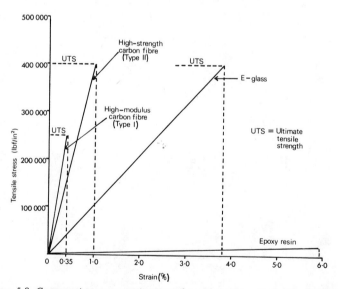

Figure 5.8. Comparative stress–strain curves for carbon fibres, E-glass, and an epoxy resin matrix material; Type I fibre possesses maximum stiffness or modulus, whilst Type II fibre possesses maximum strength

small for carbon compared with glass. Figure 5.8 indicates that, for the high-modulus carbon fibre, the maximum elongation is approximately 0·35% at break, compared with 1% for the high-strength variety and 4% for an E-glass. By comparison, an epoxy resin will normally elongate up to 5% or 6% before breaking occurs.

The important advantage of carbon fibres, therefore, is that they can withstand very high stresses without undergoing significant stretching, and hence a matrix material used with carbon is subjected to minimum stresses. This will be described more fully in Part 2 of this monograph.

### 5.3.6   OTHER PROPERTIES OF PAN-BASED CARBON FIBRES

**Table 5.5** OTHER PROPERTIES OF PAN-BASED CARBON FIBRES

| Property | High-modulus fibre (Type I) | High-strength fibre (Type II) |
|---|---|---|
| Elongation at break (%) | 0·5 | 1·0 |
| Elastic recovery (%) | 100 | 100 |
| Specific heat at 70°F(Btu lb$^{-1}$ °F$^{-1}$) | 0·17 | 0·17 |
| Specific gravity | 1·86 | 1·74 |
| Specific electrical resistance at 25°C ($\mu\Omega$ cm) | 775 | 1500 |
| Thermal expansion coefficient* (10$^{-6}$/°C) | 1·62 negative | 0·78 negative |

\* Values given for temperatures between 0°C and 50°C. At higher temperatures, namely over 230°C, the high-strength fibre expansion becomes positive.

## 5.4   FINE STRUCTURE OF PAN-BASED CARBON FIBRES

Johnson and Watt[4] and Badami, Joiner, and Jones[5] concluded that carbon fibre is made up of structural units called crystallites or fibrils, these being elongated or needle-shaped. It is believed that the fibrils may be bonded together by a secondary carbon phase. The crystallites or fibrils themselves are composed of graphite plates which can lie in random directions, except that their a-axis is less than 10° from the fibre axis in the case of high-modulus carbon fibre. It is interesting to note that Bacon and Schalamon found that a $40 \times 10^6$ lbf/in$^2$ modulus rayon-based carbon fibre consisted of graphite crystals having their a-axis oriented at 10° to the fibre axis, indicating a close fine-structure similarity between rayon-based and PAN-based carbon fibres.

Evidence for the above fine structure is provided by thin-film electron microscopy. Figure 5.9 shows an electron photomicrograph of the broken end of a carbon fibre in which the graphite plates or crystals having the correct orientation are able to cast a diffraction shadow and therefore appear in dark contrast. The fibrils can be regarded as polycrystalline and, on average, vary between 250 Å and 1000 Å in thickness. Little is known about the lengths of these

*Figure 5.9. Micrograph of a longitudinal section of a carbon fibre filament (magnification 40 000)*

fibrils, but there cannot be much doubt that they result from the breakdown of the original PAN structure. Therefore it is likely that the fibril length bears a relation to the length of the polymer chain in the precursor. Recently, Johnson and Tyson[6, 7] carried out low-angle x-ray diffraction studies of the fine structure of PAN-based carbon fibres. They found that the fibrils had widths of 50–100 Å and lengths of 1000 Å or over and postulated a schematic diagram showing the likely carbon fibre structure, which is reproduced in Figure 5.10. The structure is based on tetragonal crystals having a slight mis-orientation and enclosing sharp-edged voids or pores. The nature of the interface between adjacent crystals is not fully understood, but it is believed that a fine non-crystalline carbon exists. Johnson and Tyson are of the opinion that stretch-graphitised fibre exhibits both increased strength and increased stiffness owing to a removal of voids and also as a result of a reduction of lattice defects. This view appears to be reasonable and is supported by the work of Johnson[8], who showed that the addition of boron to carbon fibre made from PAN resulted in a higher tensile modulus and, in addition, the electrical conductivity of the fibre was increased. He attributed this to improved orientation and development of the crystallites within the fibre.

It is possible that the relationship between gauge length and breaking strength gives some evidence of the existence throughout the filament length of end-to-end joints or bonds between fibrils. It is likely that, on average, the fibrils are butt-joined together in a random fashion, so any particular cross-section of fibre will contain fibrils and some joints between fibrils. It is also probable that regions exist along the fibre length where the number of joints is greater than the average, thus creating a weak plane normal to the fibre axis. It is these areas which are likely to give rise to defects, resulting in reduced strength. All this is somewhat tentative but may, in part, explain why fibre strength is related to testing gauge length. Continued research promises to reveal more of the fine structure of carbon fibres and show the way for further improvements.

Recent work by Allen, Cooper, and Mayer[9] showed that PAN-based fibre can be marginally improved in properties by subjecting it to irradiation. Both the strength and the modulus were increased by this treatment, which appeared to cause adjacent crystal planes to bind together more firmly, probably owing to the effect of neutrons, causing interstitial bonds to be formed. Closer binding is likely to result in less slipping of the planes, giving a stiffer and stronger material. The above workers have also obtained a relatively large increase in fibre modulus by diffusing boron into the structure at graphitising temperatures, causing a recrystallisation to take place and providing a greater degree or orientation of the graphite plates; this in turn increased the stiffness of the fibre. Based on their experimental results, tensile modulus values were increased from about $61 \times 10^6$ lbf/in$^2$ up to $78 \times 10^6$ lbf/in$^2$, but strength was not significantly affected.

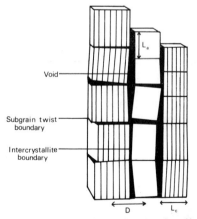

*Figure 5.10. Suggested structure of PAN-based carbon fibre. Courtesy Johnson and Tyson*[7]

The mechanical properties of PAN-based carbon fibres can be improved by stretch-graphitising, as in the case of cellulose-based fibres. Work on this has been carried out by Johnson, Marjoram, and Rose[10], and by Johnson[11]. Tensile modulus values approaching $100 \times 10^6$ lbf/in$^2$ have been obtained, together with significant increases in strength; the fibre may increase in length by almost one-third as a result of the stretching, which removes some defects and voids and improves crystal alignment.

## 5.5 TARGETS FOR IMPROVED CARBON FIBRES

There is general agreement that the stiffness or tensile modulus of a perfect graphite crystal in the $a$-direction is close to $140 \times 10^6$ lbf/in$^2$. At the same time, there is strong evidence that carbon filaments possessing modulus values at least as high as $100 \times 10^6$ lbf/in$^2$ have been made on a laboratory scale. Thus, it would seem that there is relatively little scope for large improvements in fibre modulus although one may expect to see material made available commercially with values approaching $100 \times 10^6$ lbf/in$^2$ in the not too distant future, if the user should demand these values. At present, $60-70 \times 10^6$ lbf/in$^2$ appears to be entirely adequate.

If the strengths of carbon fibres currently available are compared with the theoretical strength of a perfect graphite crystal, it will be apparent that a large difference exists. One may expect that the breaking strength should approach one-tenth of the value of the tensile modulus; thus, for a modulus of $60 \times 10^6$ lbf/in$^2$, a strength of $6 \times 10^6$ lbf/in$^2$ should, in theory, be reached. In practice, however, high-modulus fibre has a strength equal to $E/200$ and the high-strength material has a strength approximating to $E/100$, where $E$ is the modulus. It seems reasonable, therefore, to suppose that a value of $E/50$ may be attained in the foreseeable future, which would give a product with a strength of around $1 \times 10^6$ lbf/in$^2$. Such achievements will depend on a better understanding of the defects present in carbon fibre, but it must be understood that filaments exhibiting these properties will require even better surface treatment and resin systems in order to utilise fully their properties in composites.

### REFERENCES

1 BLAKELOCK, H. D., and LOVELL, D. R., 'High Modulus Reinforcing Carbon', *Proc. 24th Ann. Tech. Conf., SPI Reinforced Plastics/Composites Division,* Society of the Plastics Industry, New York (1969)
2 ROLLS-ROYCE LTD, Fr. Pat. 1 590 257 (22.5.70)

3 REYNOLDS, W. N., 'The Structure and Mechanical Properties of Carbon Fibres', Preprint of Paper 7.2, *3rd Conf. Industrial Carbons and Graphite, 1970* (to be published by Society of Chemical Industry)
4 JOHNSON, W., and WATT, W., 'Structure of High-Modulus Carbon Fibres', *Nature, Lond.,* **215** No. 5099, 384 (1967)
5 BADAMI, D. V., JOINER, J. C., and JONES, G. A., 'Microstructure Diffraction and Physical Poperties of Carbon Fibres', *J. Phys., D,* **3** No. 4, 526 (1970)
6 JOHNSON, D. J., and TYSON, C. N., 'Low Angle X-ray Diffraction and Physical Properties of Carbon Fibres', *J. Phys., D,* **3** No. 4, 526 (1970)
7 JOHNSON, D. J., and TYSON, C. N., 'The Fine Structure of Graphitised Fibre', *J. Phys., D,* **2** No. 6, 787 (1969)
8 JOHNSON, D. J., 'Direct Lattice Resolution of Layer Planes in Polyacrylonitrile-Based Carbon Fibres', *Nature, Lond.,* **226** No. 5247, 750 (1970)
9 ALLEN, S., COOPER, G. A., and MAYER, R. M., 'Carbon Fibres at High Young's Modulus', *Nature, Lond.,* **224** No. 5219, 684 (1969)
10 JOHNSON, J. W., MARJORAM, J. R., and ROSE, P. G., 'Stress Graphitisation of Polyacrylonitrile-Based Carbon Fibre', *Nature, Lond.,* **221** No. 5178, 357 (1969)
11 JOHNSON, W., 'Hot Stretching Carbon Fibres Made from Polyacrylonitrile Fibres', Preprint of Paper 7.6, *3rd Conf. Industrial Carbons and Graphite, 1970* (to be published by Society of Chemical Industry)

# Part 2

# 6

# Carbon-fibre composites: some theoretical considerations

## 6.1 INTRODUCTION

The previous chapters have discussed the development of carbon fibres having exceptionally high strength and stiffness values and methods of determining their mechanical properties. However, these materials are of little practical value unless they can be converted into a form suitable for use in engineering applications. To do this, the individual filaments must be bonded together by a matrix material, giving a composite structure whose properties depend on the properties of both the fibre and the matrix.

Considerable theoretical work has been carried out on composite structures, much of which is beyond the scope of this book, and readers wishing to acquire more detailed information should consult References 1–3 given at the end of this chapter. A more recent book[4] reports developments in the carbon-fibre composites field, with special reference to the fibre–matrix interface, and gives many additional references.

It is intended to present here only some elementary theoretical information leading to the main requirements for composites. The major part of the work carried out on composite systems has been based on glass fibre–resin composites, but the conclusions reached apply, in general, when carbon fibre is used instead of glass. It will

E

be seen that the matrix plays a vital role in the successful use of a reinforcement material, whether this be glass, carbon, or other type of fibre.

## 6.2    SIMPLE THEORY FOR LONG FIBRES

In the simplest type of composite, continuous unidirectional fibres are held in a solid matrix so that the fibres are close but not touching and both fibres and matrix are assumed to behave as perfectly elastic solids at low stresses. If such a composite is subjected to a low tensile stress, the fibres and matrix are assumed to be equally strained. Thus the composite strain $\varepsilon$ is given by

$$\varepsilon = \frac{\sigma_m}{E_m} = \frac{\sigma_f}{E_f} \tag{6.1}$$

where $E$ denotes tensile modulus, $\sigma$ denotes tensile stress, and the subscripts m and f refer to the matrix and fibres respectively.

Since the stress $\sigma$ is equal to the load $P$ per unit area $A$, and, from

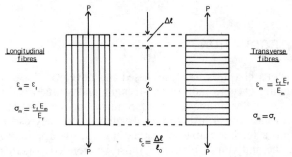

*Figure 6.1. Stress and strain relationships for composites having longitudinal and transverse fibres*

equilibrium considerations $P_m + P_f = P$ (the tensile load exerted on the composite),

$$\varepsilon = \frac{P_m}{E_m A_m} = \frac{P_f}{E_f A_f} = \frac{P}{E_m A_m + E_f A_f} \tag{6.2}$$

This equation shows that a composite will be stiff if a large number (high $A_f$) of stiff fibres (high $E_f$) is employed. It is clear that, the stiffer the composite, the smaller will be the value of the composite strain $\varepsilon$ for a given load $P$.

From equation 6.2, the following expression for the modulus $E_c$ of the composite can be obtained:

$$E_c = E_m V_m + E_f V_f \tag{6.3}$$

where $V_m = A_m/A_c$ and $V_f = A_f/A_c$ are the volume fractions of the matrix and fibres respectively, so $V_m + V_f = 1$.

Equation 6.3 may be used to predict the tensile modulus of a composite whose components are of known stiffness and present in known proportions and it implies a linear relationship between the tensile modulus of a composite and the volume fraction of the fibres employed. Further information is given in References 2 and 3.

A similar equation may be used to predict the ultimate tensile strength $\sigma_c'$ of a fibrous composite:

$$\sigma_c' = \sigma_f V_f + \sigma_m V_m \tag{6.4}$$

where $\sigma_m$ is the stress acting on the matrix at the breaking strain of the fibres. If the equal strain concept is assumed to hold up to the breaking stress $\sigma_f'$ of the fibre, then equation 6.4 may be written

$$\sigma_c' = \sigma_f' V_f + \frac{\sigma_f' E_f}{E_m}(1 - V_f) \tag{6.5}$$

The above indicates that the strength as well as the stiffness of a unidirectional composite with long fibres varies linearly with the amount (volume fraction) of the fibres it contains, since one is assuming that it is the fibres which govern the strength. Thus, a 'law of mixtures' applies and can be used as a quick and ready guide.

In the analysis, all the fibres were assumed to lie parallel to the direction of the tensile load and to be equally strong. In practice, composites do not meet these assumptions since it is not possible to align each filament perfectly and, moreover, the individual filaments do not all have the same strength. If $\theta$ is the 'orientation factor' and $\alpha$ the 'strength efficiency factor' of the fibres in any composite, equation 6.5 may be modified to

$$\sigma_c' = \sigma_f' V_f + \frac{\sigma_f' \alpha \theta E_f}{E_m}(1 - V_f) \tag{6.6}$$

Unfortunately, equation 6.6 is of little practical value since $\alpha$ and $\theta$ cannot be determined with accuracy. However, it will be seen that the orientation factor $\theta$ is governed largely by the skill in fabricating the composite; thus, where maximum performance is required, these skills must be very high indeed, as will be seen later. In the case of $\alpha$, a detailed knowledge of the fibre properties is required together with that of matrix properties and also of the mechanism of breakdown of the composite. Considering a bundle of fibres not enclosed in a matrix and subjected to a tensile stress, any fibre which breaks no longer takes its share of the load which is then shared by the remaining fibres. It can be shown that such a bundle will not be as strong as the average strength of all the individual filaments multiplied by the number of filaments in the bundle. This is because of the weakest filaments breaking first and transferring the load to the remainder. Moreover, the greater the variability in strength between filaments in the bundle, the lower is the latter's strength[5, 6]. When the fibres are embedded in a matrix and subjected to a tensile stress, the fibre breaks will occur at points of greatest weakness. However, broken fibres will still be able to support part of the tensile load owing to stress transfer through the fibre–matrix interface[3, 7, 8].

Thus, although theoretical considerations can throw much light on the behaviour of composites, they must be used with caution and practical testing of composite pieces needs to be carried out in addition.

## 6.3   USE OF SHORT FIBRES

If a short fibre is embedded in a matrix which is subjected to a tensile stress, it will take up the stress to an extent related to the ability of the matrix to sustain shear deformation. The fibre stress will build up from zero at the fibre ends and, if the fibre is long enough, reach a certain constant value. As the stress on the composite is progressively increased, this value will be raised to a point where either the fibre breaks in tension, or the matrix fails by shear

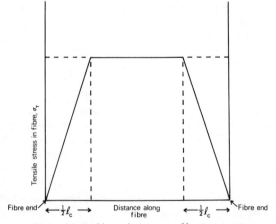

*Figure 6.2. Build-up of stress on a fibre in a matrix*

flow, or there is a rupture of the fibre–matrix bond. If the fibre in question is shorter than a certain 'critical length', the stress will not be able to build up to a value sufficient to cause fibre fracture and shear failure will occur instead[3, 7, 8]. The concept of a minimum fibre length arises because the actual ends of the fibre cannot support a tensile load. The load builds up from the ends of the fibre, as shown in Figure 6.2, and the critical length $l_c$ is given by the expression

$$l_c = \frac{D\sigma_T}{4\sigma_S} \tag{6.7}$$

where $D$ is the fibre diameter, $\sigma_T$ is the tensile stress in the fibre, and $\sigma_S$ is the shear strength of the fibre–matrix interface.

It will be seen that the critical fibre length is greater with a weak matrix bond than with a strong one since the latter allows a more rapid build-up of load along the fibre length.

## 6.4    MECHANISM OF COMPOSITE BREAKDOWN

From the foregoing it will be seen that two elements of uncertainty are introduced into the calculation of an effective value of $\alpha$ in the composite:

1. The stress-bearing 'contribution' of the broken fibres.
2. The pattern of fibre breaks that occur and eventually cause the breakdown of the composite.

It is important to know whether the fibre breaks occur randomly, or whether there is a pattern of breaks indicative of local stress transfer, voids, or flaws in the matrix.

There are differences of opinion about the actual mechanism of composite breakdown, which hinge on the question of random versus local breaking of filaments. A possible mechanism of composite failure was considered by Parratt[9], who suggested that breakdown took place as a result of random-sized fractures when the fibres had been shortened to the extent that they were unable to carry the applied stress. Failure would then occur by shear, either of the matrix material or of the fibre–matrix bond. To obtain the actual composite breaking strength, Parratt equated the maximum stress that could be carried in a glass–resin composite by discontinuous fibres of different lengths with actual strengths of free fibre bundles of the same lengths. From his equation, he obtained a value of 400:1 for the aspect ratio of E-glass fibres in a resin composite.

A different mechanism suggested by Rosen[8] implies that the breakdown is caused by accumulation of defects in one region of the composite. Failure takes place when a large number of breaks occurs in this one area and the load applied to the composite cannot be transferred across some plane perpendicular to the direction of the load. Thus, final failure occurs across this plane.

Both of these mechanisms assume that a single fibre break will not cause more than a local stress discontinuity. In practice, this may not be the case. If the matrix is not ductile and the fibre–matrix bond is weak, interfacial separation of fibre from the matrix may spread laterally along the composite from the fibre break, rendering the fibre concerned incapable of bearing the stress. If this happens many times, it will undoubtedly hasten the complete breakdown of the composite. A more serious situation may arise if the fibre–matrix bond is strong, as the fracture toughness of the matrix may not be sufficiently high to prevent a crack spreading laterally across the composite from the point of fracture, perhaps causing catastrophic failure[8, 10, 11]. To avoid the risk of this, it has been suggested that the fibre–matrix bond should not be too strong if a non-ductile matrix, such as a very hard resin or a metal, is employed; under these circumstances, the lateral cracks will tend not to occur or will be deflected along the fibre–matrix interfaces when they do[3, 10, 12].

Arguments have also been advanced in favour of making the fibre–resin bond as strong as possible on the grounds that stress discontinuities caused by fibre breaks will not then be transferred to other regions of the composites by progressive interfacial break-down, thereby causing further breaks at other points of fibre weakness. A further advantage is that the shear strength of the composite will be kept high[13–15].

In the absence of a satisfactory theoretical treatment, the choice between these two alternatives for a given composite will depend largely upon the actual mode of breakdown that occurs, and how it is affected by varying the strength of the fibre–matrix bond. In general, breakdown in a carbon fibre–resin composite will be expected to occur as Rosen suggests; that is, by accumulation of fibre fractures to the point of causing a plane of weakness in the composite. Cracking the resin as a result of a single fibre fracture, though structurally undesirable, need not initiate breakdown. Based on these arguments, the following should be ensured:

1. A stiff matrix material so that stress discontinuities at fibre breaks are as localised as possible.

2. A strong fibre–resin bond and adequate fibre length so that stress discontinuities are not propagated along the fibres to other points of fibre weakness.
3. A high volume fraction of stiff fibres so that the strain and consequent shear stresses on the matrix are kept as low as possible.

It is suggested that these be taken as a guide only. The breakdown mechanism of a composite is very complex indeed, and the theories of breakdown proposed have been shown to be of strictly limited value in practice owing to many factors which are difficult to quantify. Thus, in glass fibre the effect of surface damage and abrasion can render calculations invalid. Equations 6.5 and 6.6 serve as only a very rough guide but may be of value to indicate actual progress against the theoretical value.

## 6.5 THE MATRIX–FIBRE BOND

It has been shown that the simple theory for determination of composite strength does not agree with the results obtained in practice. The presence of defects in the fibres themselves leads to breaks and stress concentrations in the composite; crossed fibres and void areas also cause stress concentrations and result in departure from simple theory. One of the most important reasons for this lack of agreement, however, is that two contradictory assumptions have been made:

1. That the fibres and resin matrix are intimately bonded together in the composite.
2. That the fibre and resin phases behave as separate entities, each possessing its own stress–strain behaviour.

The microstructure of resins themselves indicates that many possess a 'micelle' structure, in which relatively high stiffness regions are embedded in a medium of lower stiffness. The micelles are about 800 Å in diameter and, in an unreinforced resin, they may move and even assist the propagation of a crack. In a reinforced resin system, however, the reverse may be true since the micelles close to the fibre interphase layers behave differently from those further away: the former will have more limited movement because of the close proximity of the fibre surface. Thus, in practice, the resin matrix does not behave like a homogeneous medium.

If the bonding between the fibre surface and the resin is intimate, even under conditions of no external loading there may be internal stresses set up in the composite owing to differences in expansion coefficients between the two components. Thus, the contraction of the resin after heat treatment (curing operation) will create a compressive stress on the fibre and a tensile stress on the matrix, which, from elementary mechanical considerations is given by[5]

$$\sigma_m = E_m \, \Delta T(a_m - a_f) \qquad (6.8)$$

where $\Delta T$ is the fall in temperature from the curing operation and $E_m$ is the tensile modulus of the matrix; $a_m$ and $a_f$ are the thermal expansion coefficients of the matrix and the fibre reinforcement respectively. In many composites where the matrix is a metal, the cooling stress is sufficient to cause considerable damage. Equation 6.8 may be used as a guide to the compatibility of matrix and fibre: $\sigma_m$ may be as high as 100 000 lbf/in$^2$ for a metal matrix and this value represents a severe loss in composite properties[5]. However, for resins the value lies much closer to 1000 lbf/in$^2$. In practice, equation 6.8 is only a guide since the actual stresses depend on the volume fraction of the fibre, in addition to the properties of fibre and matrix. Moreover, if the volume fraction of the fibre is low, buckling and distortion can take place to accommodate the stress. For high fractions, photostress analysis has shown the presence of very complex stress patterns[1, 16].

One consequence of the tensile stresses created in the resin matrix is that the micelle-type structures open up and become more porous so water is more readily absorbed. Epoxy and polyester resins behave in this manner. Thus, a cast non-reinforced resin absorbs less water than one which has been subjected to tensile stresses owing to reinforcement, and this explains why one particular glass–epoxy composite tested by Eakins[16] absorbed up to 0·6% of its own weight of water when not subjected to any external stress, but only 0·2% under a compressive loading of 7000 lbf/in$^2$.

The work of Eakins is noteworthy. He studied the correlation between the shear stress in resin castings and the shear strength of glass fibre–resin composites as measured by: (a) short-beam flexural testing, and (b) compressive testing. In both cases he reported a fairly good correlation except where the compressive test samples contained structural voids. His results suggest that, in the short-beam flexural test, breakdown of glass fibre–resin composites occurs owing to shear failure of the resin in regions beyond the interphase layer. The work also established that the compressive strength is dependent on the resistance to shear and is lowered by the presence of voids in the composite structure, these presumably initiating premature delamination. The work of Fried, Kaminetsky, and

Silvergleit[17] has also led to similar conclusions. However, in order to arrive at a complete understanding of the structure of a carbon–resin system and of the failure mechanism, much further work will have to be carried out.

## 6.6  LAMINAR COMPOSITES

The vast majority of practical composite structures employ the reinforcement materials in more than one direction in order to meet structural requirements. Composites which are not unidirectional are more complex in behaviour than those considered up to now, and the theory becomes even more limited in value.

By considering the effect of a low stress acting at an angle $\phi$ to the fibre direction, strains occurring in the fibre–matrix may be calculated in the appropriate perpendicular directions without difficulty. But if the stress is sufficiently high to cause significant non-elastic deformation of the matrix, the whole situation becomes more complex. Morley[12] describes the limitations imposed upon the applied stress for four classes of deformation:

1. Matrix flow parallel to the fibres causing fibre failure; failure stress

$$\sigma_d = \sigma_c \cos^2 \phi \qquad (6.9)$$

2. Matrix shear failure (shear debonding failure); shear failure stress

$$\tau_u = \sigma_c \sin \phi \cos \phi \qquad (6.10)$$

3. Matrix flow perpendicular to fibre direction; transverse flow stress

$$\sigma_t = \sigma_c \sin^2 \phi \qquad (6.11)$$

4. Transverse tensile debonding failure; failure stress

$$\sigma_{dt} = \sigma_c \sin \phi \qquad (6.12)$$

Throughout these equations, $\sigma_c$ is the stress acting on the composite.

In the case of epoxy or polyester resins which have no appreciable flow properties, the only relevant modes of failure are those of matrix shear—implying shear failure of the resin or the fibre–resin bond—and transverse debonding. Thus, the maximum stress that may be exerted on the composite at an angle $\phi$ to the fibre direction will be given by

$$\sigma_{max} = 2\tau_u \operatorname{cosec} 2\phi \qquad (6.13)$$

for shear failure, and

$$\sigma_{max} = \sigma_{dt} \operatorname{cosec} \phi \qquad (6.14)$$

for tensile debonding failure.

Experiments carried out by Broutman[14] on single glass fibres in resin matrices have established values of 3250 lbf/in$^2$ and 1550 lbf/in$^2$ respectively for shear and tensile debonding strengths of glass fibre–epoxy resin bonds using reliable coupling agents. In actual composites, however, the presence of neighbouring fibres modifies these values. Broutman has shown that, for composites of from 40% to 80% fibre volume fraction, the theoretically calculated shear and tensile debonding failure stresses were 9620 lbf/in$^2$ and 5180 lbf/in$^2$ respectively. Application of equations 6.13 and 6.14 to these figures indicates that a unidirectional composite would always fail by tensile debonding and, at an angle of only 10° to the fibre axis, would exhibit a tensile strength of only 30000 lbf/in$^2$. Such a result indicates the vital importance of having a multidirectional reinforcement in practical glass-fibre composites to prevent the occurrence of tensile debonding. The same is true for carbon-fibre composites.

With bidirectional fibre composites or filament-wound structures, one of the most difficult problems is that of 'strain magnification', which takes place in a particular plane along a direction perpendicular to the fibre axes in that plane as a result of an external stress in that direction. It is characterised by a higher strain occurring in the plane than would be predicted from elementary composite theory[18, 19]. In bidirectional glass fibre–epoxy resin composites, this phenomenon gives rise to 'crazing' or cracking of the resin in the plane concerned at stresses which correspond to about 25% of the ultimate tensile strength corresponding to a strain of 1·7% when the composite undergoes fatigue testing in tension or flexure. The crazing of the resin matrix is not serious in itself since the fibres support the load, but it allows water to enter the resin and reach the glass-fibre surfaces and subsequently cause loss of strength.

Carbon fibres, as will be shown later, do not suffer from the above effect to anything like the same extent, but the problem is still inherent in any type of fibre–matrix system in which the fibres lie in more than one direction.

## 6.7   CONCLUSIONS

The theoretical work described above and based largely on glass fibre–resin systems has shown that, in order to realise the maximum benefit from fibre reinforcement, a number of requirements must be fulfilled.

1. The fibre content should be as high as possible since the fibres are the source of strength. In practice, up to 70% by volume of fibre can be achieved with care.

2. The alignment and direction of the fibres is of paramount importance. The proportion of fibres in any direction determines the properties in that direction. Each of the fibres should be in the correct position to ensure maximum contribution to load bearing. Overall alignment determines the composite properties in the desired direction, and fibres should lie side by side and close together. Twist in the reinforcement, even though it may involve only a small percentage of fibres, is detrimental as twisted fibres do not contribute fully but do give uneven resin distribution.

3. Fibres should be wetted out fully by the resin to ensure good contact and bonding at the interface so that the load may be transferred from one fibre to another throughout the composite.

4. Good bond strength at the resin–fibre interface is essential for load transfer in the composite.

5. Fibre quality is important. Freedom from defects such as an excessive number of broken filaments is essential, although the effect is minimised when good adhesion occurs between fibre and resin, the latter transferring the load across the break. Kinks and twists in the fibres, however, produce stress concentrations, etc., as described in requirement 2.

6. The matrix must be compatible with the fibre and should not debond or crack when the fibre undergoes maximum strain at full load. Fatigue properties, where stresses undergo many reversals, depend largely on such compatibility. In composite structures containing carbon fibre, the strain of the fibre at full load is one-third or less of that of glass fibre and the tendency for debonding is reduced accordingly.

7. In composite bodies made up of dissimilar fibres such as glass and carbon particular care needs to be given to the stress compatibility of the whole. It stands to reason that glass fibres and carbon fibres cannot be laid parallel to each other since the carbon would take all the load. However, they may be laid at different angles in many applications.

Full answers to the many requirements for optimum composite performance will be forthcoming only when better understanding resulting from further work has been achieved. Such work will have to be carried out on a wide range of carbon-fibre composites, tested under a variety of conditions.

REFERENCES

1 HOLLIDAY, L., (ed.) *Composite Materials*, Elsevier, Amsterdam (1966)
2 HOLISTER, G. S., and THOMAS, C., *Fibre Reinforced Materials*, Elsevier, Amsterdam (1966)

3  KELLY, A., and DAVIS, C. J., 'The Principles of Fibre-Reinforced Metals', *Metall. Rev.,* **10** No. 37, 1 (1965)
4  AMERICAN SOCIETY FOR TESTING AND MATERIALS, *Interfaces in Composites*, ASTM Special Technical Publication 452 (1969)
5  SCALA, E., 'The Design and Performance of Fibres and Composites', *Fiber Composite Metals*, American Society for Metals, Cleveland (1964)
6  FOSTER, B. K., and BEER, F. J., 'Fibre-Reinforcement; The Mechanics of Fibre Reinforced Materials', *Eng. Elect. J.,* **21** No. 4, 18 (1966)
7  COTTRELL, A. H., and KELLY, A., 'The Design of Strong Materials', *Endeavour,* **25** No. 94, 27 (1966)
8  ROSEN, B. W., *Mechanics of Composite Strengthening*, Space Sciences Laboratory, Mechanics Section, Report R64SD80, General Electric Co., King of Prussia, Pennsylvania (1964).
9  PARRATT, N. J., 'Defects in Glass Fibres and Their Effect on the Strength of Plastic Mouldings', *Rubb. Plast. Age,* **41** No. 3, 263 (1960)
10  COOPER, G. A., 'The Potential of Fibre Reinforcement', *Metals & Mater.,* **1** No. 3, 109 (1967)
11  US MATERIALS ADVISORY BOARD, *Micromechanics of Fibrous Composites*, Report MAB-207-M (1965)
12  MORLEY, J. G., 'Fibre Reinforced Metals', *Sci. J.,* **2** No. 11, 42 (1966)
13  REED, J. V., 'Plastics Composites', *Chem. Process Engng,* **48** No. 2, 63 (1967)
14  BROUTMAN, L. J., 'Glass Resin Joint Strengths and Their Effect on Failure Mechanisms in Reinforced Plastics', *Polym. Sci.,* **6**, No. 3, 232 (1966)
15  DOW, N. F., *Study of the Stresses Near a Discontinuity of a Filament Reinforced Composite Metal*, Report TIS R63SD61, General Electric Co., King of Prussia, Pennsylvania (1963)
16  EAKINS, W. J., *Initiation of Failure Mechanisms in Glass-Resin Composites*, NASA Contractor Report NASA-CR-518, National Aeronautics and Space Administration, Washington (1966).
17  FRIED, N., KAMINETSKY, J., and SILVERGLEIT, M., 'The Effect of Deep Submergence Operational Conditions on Filament Wound Plastics' *Proc. 21st Ann. Tech. Conf. SPI Reinforced Plastics Division,* Society of Plastics Industry, New York (1966)
18  MCGARRY, F. J., *Plast. Technol.,* **5**, 44 (1969)
19  SCHWARTZ, H. S., 'Principles of Fibre Reinforced Structures in Plastics Components', *SAMPE Jnl,* **2** No. 4, 19 (1966)

# 7

# Fabrication of carbon-fibre composites

## 7.1  INTRODUCTION

The simple theory presented in the previous chapter indicates that the strength and modulus of a composite structure are proportional to the volume fraction of the reinforcing fibres, other factors remaining constant. Thus, the greater their volume, the higher the strength and stiffness of the resulting composite. The discussion in Chapter 6 also highlights the fact that theoretical calculations may differ from experimental results owing to a number of factors which are difficult to quantify, namely:

1. The orientation of fibres in a practical composite, which may depart from the ideal.
2. The length–diameter ratio of the fibres, which may not be known.
3. The presence of rogue filaments having appreciably lower than average strength.
4. The degree of bonding between the fibres and the resin matrix, which cannot be easily quantified.

Fabrication techniques have an important bearing on these factors and must ensure that all the fibres make the maximum contribution to the strength of the composite if maximum performance is to be obtained. Whereas glass fibre–resin composites have been produced for many years, relatively little experience exists for carbon fibre–resin systems. Moreover, carbon fibres possess different handling characteristics from glass, necessitating the development of new methods. In addition, the object of using carbon fibre in composites is to obtain the best possible mechanical properties, which emphasises the need for advanced fabrication techniques.

Much work is currently being carried out on the fabrication and subsequent evaluation of carbon-fibre reinforced composite systems. This chapter considers the most important methods of fabrication used at the present time, but it must be remembered that these are being continually improved and new techniques are under development. As already stated, the resin component plays a key role in the use of carbon fibres in high-quality composite structures, but resins will be considered in detail only in the next chapter.

## 7.2 SIMPLE MOULDING TECHNIQUES

The simplest method of making a composite is to take individual fibre tows, soak them in resin, and lay them side by side in a mould in order to produce a unidirectional article such as a rod, strip, or plate. Pressure is applied to the top plunger of the mould, together with heat to cure partially the resin. During compaction, excess resin is squeezed out and the filaments remain approximately parallel. After cooling, the mould may be opened to release the composite, which is then fully cured in the normal manner.

This technique can produce good composites if great care is taken. However, it does not represent a viable production method and often fails to give consistent composite performance for the following reasons:

1. Excess resin is nearly always contained within the composite, and its distribution is far from uniform.
2. The fibres do not remain fully aligned during the compaction operation.
3. There is a tendency for air to become trapped in the mould and to be retained in the composite.

A further serious drawback to this method is the difficulty of aligning fibres in more than one direction. More sophisiticated methods of fabrication have therefore been developed and will be described in the following sections.

## 7.3   FABRICATION OF COMPOSITES USING THE PRE-IMPREGNATION TECHNIQUE

The simple method of moulding a composite bar described in Section 7.2 is of little value for the majority of components where the optimum properties are desired. One of the most important features of a good composite material is that the fibres have a very good degree of alignment in the required directions. To achieve this aim, the technique of 'pre-pregging' has been developed in which the fibres are pre-impregnated with the correct amount of resin and aligned in the form of sheets or tapes which can be handled easily.

### 7.3.1   PRODUCTION OF PRE-PREG SHEET AND TAPE

Filaments or tows are first coated with the resin to be used as the matrix and are aligned parallel to each other on a suitable non-stick surface. Siliconised paper or plastic sheet or film is frequently used as it can be peeled off at a later stage. Apart from parallel alignment, the tows must be evenly spaced, and, when this has been achieved, a second sheet of siliconised paper is laid on the top of the tows, as shown in Figure 7.1(a). It will be seen that the resin-coated tows have a circular cross-section at this stage.

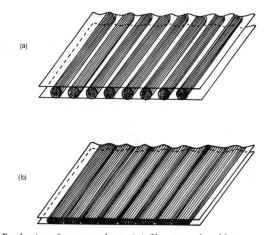

*Figure 7.1. Production of pre-preg sheet: (a) fibre tows placed between two sheets of plastic film or siliconised paper (tows are first coated with resin and are carefully spaced and aligned parallel to each other; the plastic film may be peeled off at a later stage); (b) fibre tows after being rolled between the sheets of plastic (the tows have spread outwards and adjacent tows now meet; fibre alignment is preserved and the action of rolling removes trapped air and ensures an even thickness; heat is usually applied to assist the rolling operation)*

The tows are then rolled between the two sheets to flatten them so that the filaments in adjacent tows meet, filling the gaps between them. A tow of circular section may have a diameter of 2 mm but, after rolling, a spread up to 6 mm or even 12 mm may occur. The amount of rolling controls the thickness of the sheet, but further control is obtained by the choice of initial gap between tows. Figure 7.1(b) shows the rolled tows and the spreading of filaments across the sheets. During the rolling, heat is frequently applied to provide greater mobility in the resin and assist the flattening operation; in addition, the heat partially cures the resin so that the resulting material is flexible but not too sticky. At this stage, the papers can be peeled back revealing a pre-preg sheet in which all the filaments are aligned parallel. Moreover, the sheet can be made to a controlled thickness—say 0·010 in or even 0·005 in—and the action of rolling eliminates any resin-rich areas and air bubbles. It is possible to produce sheets with at least 55–60% volume fraction of carbon fibre as a matter of routine. Such a material is much more suitable for the manufacture of high-quality composite pieces, particularly where good reproducibility is essential. An important feature of pre-preg is that its weight per unit area may be determined with accuracy and this assists the subsequent moulding operation.

Pre-preg sheets will naturally have a shelf-life dictated by the resin used, and this may vary from days to months at room temperature. Frequently solvents are added to the resin to improve its flow characteristics and the wetting-out of individual filaments. Presence of a solvent influences the shelf-life of a resin, but solvents must be removed during the subsequent composite manufacture if optimum properties are to be obtained. To increase the shelf-life of the resin, pre-preg is frequently stored under refrigeration.

The manufacture of high-quality pre-preg sheet calls for skill and expertise. The primary requirements in pre-preg manufacture are a high fibre content and a very high degree of uniformity with virtually no gaps between the expanded tows. Equally important is that all the filaments are wetted out by the resin; whereas this may be straightforward with, say, an epoxy resin diluted with solvent, other resins such as high-temperature polyimides present greater handling problems. The quality of the carbon fibre in terms of appearance and handling characteristics also plays a major role. Fibre itself should be as free from waves and kinks as possible since these can never be completely eliminated in the pre-preg tape or sheet. Moreover, the fibre has to be free from twist so that the filaments in the tow can spread out to the desired extent. An excessive number of loose filaments, or hairiness in the fibre, is detrimental since these fail to align themselves and do not contribute fully to the composite strength. In addition, loose filaments are often the cause of local resin concentrations, which is also a drawback.

### 7.3.2    FABRICATION OF UNIDIRECTIONAL COMPOSITE FROM PRE-PREG

A sufficient number of the pre-preg sheets is loaded into the mould to give the required 'fill'. The mould is then heated to an appropriate temperature, and pressure is applied to the top plunger so that the whole load is compacted, this operation also serving to part-cure the resin. It is normal to use a 'leaky mould' to allow excess resin to run out. If the pre-preg sheet is of high quality, the amount of excess resin will be virtually nil, owing to its high fibre content. Figure 7.2 shows the moulding operation. It will be seen that a

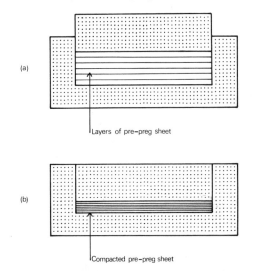

(a)

Layers of pre—preg sheet

(b)

Compacted pre—preg sheet

*Figure 7.2. Moulding of a unidirectional composite from pre-preg: (a) loaded mould before compaction; (b) mould after compaction (the top plunger has been pressed down until level with the mould, giving a composite of known thickness and density)*

convenient way of working is to load a given number of pre-preg sheets and hence a known weight of fibre and resin into the mould, followed by pressing until the top plunger reaches a closed position. In this way, the dimensons, density, and fibre content of the moulded piece can be held to close limits. The main advantage of this method is that it allows fibre alignment to be preserved and at the same time permits the maximum quantity of fibre to be uniformly introduced. After cooling, the mould is stripped, the component is removed, and final curing of the resin is allowed to take place at the appropriate temperature. Alternatively, the full curing operation may be carried out before the mould is opened. To facilitate release from the mould, it is necessary to use a release agent which is sprayed or painted on to the mould surfaces before use. Two release agents in wide use are Mold-Wiz and Releasil 14.

A development of the above moulding technique has taken place in the aircraft industry, where the size of components prohibits the use of conventional steel moulds. In the case of a large structural part—a box-section wing-beam, for example—the technique is to lay tape on to a profiled surface which has the contours required for the under surface. The tape, say 3 in wide and consisting of unidirectional pre-preg, is laid on to this former by means of a tape-laying machine which may be numerically controlled so that the fibre tape is precisely placed and oriented. Moreover, such a machine allows the different layers to be built up as required and with the requisite number of tapes in the chosen direction. On completion of the pre-preg tape-laying operation, a plastic bag is placed over the whole and air is pumped out, a process which compacts the tapes as a result of atmospheric pressure. Further pumping removes solvents from the resin, together with any entrapped air. The whole is then heated in an autoclave to cure the resin completely under increased pressure, yielding a finished component to any desired size and requiring the absolute minimum of machining. Such fabrication methods are revolutionising the use of fibre-reinforced composites.

### 7.3.3 METHOD OF FABRICATION OF A BIDIRECTIONAL COMPOSITE

The method used is essentially the same as for the unidirectional composite. The key feature of the operation is once again the use of pre-preg sheets or tapes, which are laid one over the other so that the fibres run in the required directions. Many layers or plies may be used to establish the necessary component thickness, and the number of layers with fibre running in a particular direction is governed by the characteristics required in the composite. Figures 7.3(a) and 7.3(b) respectively show an example of a bidirectional pre-preg and a multidirectional pre-preg sheet. The processing and curing stages follow in exactly the same manner as for the unidirectional material.

It will be seen that the fabricator has an infinite choice of fibre angles and a large choice of the number of layers containing fibres in any particular direction in a single plane. He can thus 'tailor make' a composite with any desired properties in the one plane.

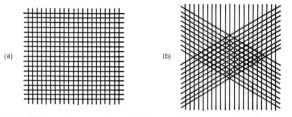

*Figure 7.3. Mouldings of composites: (a) layers of pre-preg with fibres running at right angles to each other; (b) layers of pre-preg with fibres running at 60° to each other*

Where strength is needed in a second plane, the normal method is to bend the layers into that plane, as in a simple box, for example. However, if the reinforcement is required in a truly three-dimensional sense, normal pre-pregging cannot be readily adapted and the use of chopped fibre or even a woven form of fibre may be necessary: these methods are described later.

### 7.3.4 USE OF RANDOM-MAT TECHNIQUES

The random-mat technique is one which has found widespread use in the glass-fibre field and involves the random laying of glass fibre to form a mat, which is then used to reinforce resins. A random mat with uncured resin is flexible and can be bent to the required shape—as, for example, in the construction of a boat hull. In effect, the random orientation of the fibres confers similar mechanical properties in all directions in the plane of the sheet, and, since it can be carried out at a low cost, it has found much use with glass fibre. There is no reason why it could not be used with carbon fibre, and it would be cheaper than the use of carefully aligned pre-pregs; however, the random orientation would result in numerous resin-rich areas and inferior strength. With a high-performance material like carbon fibre this clearly is not the best method of utilisation, and its use is therefore likely to be limited.

### 7.3.5 USE OF CHOPPED-STRAND MAT TECHNIQUES

This technique is very similar to the previous method except that the fibre is chopped to, say, $\frac{1}{2}$ in lengths so that the fibre tow remains intact, and the chopped strands are spread out in a sheet and coated with resin. The properties obtained are probably worse than those obtained with the random-mat technique, but the use of chopped fibre may be cheaper. A chopped-strand mat pre-preg sheet is also flexible and may be bent to shape as required.

## 7.3.6   USE OF CHOPPED STRANDS AND MOULDING COMPOUNDS

Glass fibre is used on a relatively large scale for the reinforcement of moulding components. The glass in the form of a yarn or tow is sized so that the individual filaments are bonded together, after which a chopping operation reduces them to a length of say $\frac{1}{4}$ in or $\frac{1}{2}$ in; the size also improves the bonding of the fibre with the moulding compound. The glass is thus in the form of short rods, each containing many filaments, and is in a suitable condition for loading into a resin or plastic moulding compound. A mixing operation is carried out to introduce the glass but, owing to the high viscosity of the mixture, which has the consistency of dough and has given rise to the term 'dough moulding compound', it is difficult to obtain a fibre content much higher than 20–25% by volume. Nevertheless, the presence of this amount of glass reinforcement results in a much stiffer and stronger moulded article; moreover, the latter has a greater dimensional stability than the unreinforced material. A notable feature of this process is that the fibres tend to distribute themselves in all directions to give three-dimensional reinforcement, and the presence of the glass has very little effect on the behaviour of the compound during moulding.

High-modulus or high-strength carbon fibres can be used in the same way as glass fibres. The fibre tows are first of all sized with a suitable material, which may be a resin, to keep the tows intact during chopping and subsequent mixing with the moulding compound. Failure to bond the tows in this way results in the single filaments being broken excessively at the mixing stage; with carbon the breakage is greater than with glass owing to its higher modulus and brittleness. Again the loading is limited to, say, 25% by volume, and even this is difficult to achieve. Because of the relatively low fibre content, the mechanical properties of moulded articles are inferior by comparison with those produced by the pre-preg method. Use of mobile thermosetting resins may assist the fabricator to achieve higher fibre loadings but introduces difficulties at the moulding stage. Thermoplastics such as nylon or PTFE are difficult to reinforce with carbon fibre and do not acquire very promising mechanical properties, as will be seen later.

## 7.3.7   WEAVING OF FIBRE

Handling of carbon fibre during the conversion into a moulded article or mechanical structure may be facilitated by the use of fibre in a woven form. A woven fabric can be coated with resin immediately before the laying-up operation to build the required component shape, and hence the problem of resin shelf-life does not arise.

However, high-modulus carbon fibre is difficult to weave owing to the brittleness of the individual filaments. The high-strength fibre, on the other hand, may be woven successfully provided great care is taken during the weaving operation, which necessitates the application of sized coatings to the fibre prior to weaving. An alternative is to take partially processed carbon fibre, as typified by oxidised PAN, which possesses much easier handling characteristics than fully converted fibre. It is therefore possible to weave these materials and carry out the subsequent carbon fibre processing afterwards.

At present, work is being undertaken to investigate the potential of the weaving method. A logical extension of the technique is to produce a woven form in which carbon fibres are aligned in one direction and other materials such as glass are aligned in the other direction; clearly this opens up many possibilities but increases the complexity of the overall operation. Nevertheless, success has been achieved with the weaving of the high-strength form of carbon fibre, this being used as the weft and a nylon or other thread forming the cross stitch. Such a material can be rolled up like cloth, which facilitates handling.

## 7.4 USE OF FILAMENT WINDING METHODS

Filament winding methods are used for the production of shapes such as bottles, cylinders, and tubes. Choice of winding method allows the fabricator to place fibres in the directions required to give the desired properties. The method has been used extensively with glass fibre and is now being used with carbon fibre in the form of long lengths. Since the fibres are placed in a precise manner, the method allows maximum performance to be derived from them, and the technique is akin to the use of aligned pre-preg sheets and tapes. By the choice of suitable mandrels, the filament winding method can produce a wide variety of components which may include the incorporation of metal end mountings, screwed connectors, etc. Frequently the mandrels have to be made collapsible to allow removal after winding and pre-curing. Final curing can then be completed.

Based on the development in the glass-fibre industry, there are four main methods available for winding carbon fibre. The first method uses a lathe or similar machine in which parallel winding can be carried out or in which cross-winding patterns may be utilised. In either case, a reciprocating feed or guide is fitted to the lathe and made to move in the appropriate manner. Such a machine is suitable for the manufacture of filament-wound cylindrical objects. The second technique involves the use of an orbital winder which can produce a greater variety of patterns and gives a longitudinal wind. In this process, the thread guide moves around the mandrel in an orbital fashion. In the third method, which is similar to the previous one, the mandrel rotates in such a way as to produce the desired pattern which may include longitudinal, helical, or circumferential winding. The last method is sometimes known as the 'whirling arm' method and is the most versatile since any combination of winding patterns may be generated and the machine used with it has been developed by combining the attributes of the earlier types.

In order that optimum composite properties may be realised from filament winding techniques, fibre appearance or 'cosmetic' qualities are of considerable importance. The fibre should be free from twist and, if possible, any false twist, but in practice the latter is always present to some degree. Absence of twist allows controlled spreading of the tow to be achieved; this is important as it determines the effectiveness of filament placing. Minor waves in the fibre, whilst undesirable, are not particularly detrimental since the tension of winding eliminates them. Excessive hairiness or fuzziness in the fibre has to be avoided since stray filaments become out of line and thus ineffective; moreover, such filaments prevent accurate control of resin content and reduce the compactness of the winding.

The handling of carbon fibre during filament winding requires further development. With glass fibre, special techniques have been evolved over the years to prevent fibre damage as the winding takes place under controlled tension. Coatings are frequently applied to lubricate the surfaces of the glass and to provide protection. It is probable that a similar approach will have to be made to carbon fibre to allow an even application of the material under tension, as in glass filament winding. It should be noted that, with a high-modulus material, the effect of tension variation is more significant than with glass fibre since, with the latter, a higher strain is exhibited under load and hence the material can accommodate variations to a greater degree.

The development, particularly in the USA and the UK, of high-temperature resins of the polyimide type has raised a number of fabrication problems. In filament winding, the problem is to wet out the resin effectively and yet prevent an excessive resin content. A technique developed by Brown and Marshall of the US Air Force Materials Laboratories[1] uses electrical heating of the fibre in order to reduce the polyimide resin viscosity and so obtain better flow and wetting at the surface. The technique may be applied to any resin to obtain improved control of resin content.

For the fabrication of large or complex shapes by the filament winding process, specially designed machines are usually employed. This is particularly true in the US aerospace industry where such machines have been designed to allow precise placing of the fibre so that any desired strength pattern can be produced. Use of these machines allows production runs to be made with the certain knowledge that each component will have the same performance as the others; they form a parallel with the uniaxial tape-laying machines which have been mentioned earlier.

Hardesty[2] has described a new technique which extends the production capabilities of the filament winding process. The method consists of carrying out continuous winding and concurrent curing of the resin matrix so that tubes or sections can be fabricated on a truly continuous basis. This method has one or two features in common with the pultrusion process described below.

## 7.5   PULTRUSION METHOD

The specialised 'pultrusion' process has been developed for use with glass fibre and is specifically related to the production of beams, tubes, and channel sections. Basically, it consists of drawing glass-fibre rovings through a resin bath, followed by passage through a bush of the required shape to remove the excess resin and any entrapped air. The drawn material is then passed continuously through an oven to harden the resin. The process was developed in the 1950s in the USA, where the name 'pultrusion' was coined to describe it. It is believed that it can be adapted to employ carbon fibres to give sections of higher mechanical performance.

There are at least two variants of the basic pultrusion process. In one, a die is used to give precise dimensions to the drawn material. The process is arranged so that, after passing through the coarse bush, the material has to move through a tube die which confers the exact shape and dimensions to the section and at the same time cures the resin by microwave heating. The latter is very fast and provides uniform heating throughout the section; with a well-designed unit, full hardening of the resin can be carried out in under 24 in length of tube die*. In a second version, an externally heated split die is used for curing purposes and for determining the dimensions and shape of the extruded material. Since curing usually

* Glastrusions Inc. in the USA manufacture such equipment.

takes place more slowly, the process may have to be carried out in discrete steps, a section remaining stationary in the die during the cure and being followed by the movement of a new section into the split die. In effect, a continuous length may be produced from a number of separately cured sections*.

Carbon fibre is likely to play a major part in further development of pultrusion techniques, as either a full or a partial substitute for the glass fibre to give improved mechanical properties. Fibre for use in this type of process is likely to take the form of long lengths, but the use of some coarse woven form may also be possible. If high fibre contents and optimum properties are to be achieved, much evaluation work will have to be carried out, one important aspect being the control of resin flow and wetting.

REFERENCES

1 US AIR FORCE. US Pat 3 339 252 (27.8.68)
2 HARDESTY, E. E., 'Continuous Filament Winding with Continuous Curing', *Today's Thermosetting Plastics* (*Proc. Southern California Section Regional Tech. Conf., Manhatten Beach*), Society of Plastics Engineers, Los Angeles (1969)

* In the USA firms such as Goldsworthy Engineering Inc. make equipment for the semicontinuous pultrusion process while, in the UK, GEC–English Electric are involved. The latter make a range of components under the trade name Extren.

# 8

# Resins and other matrix materials for carbon-fibre composites

## 8.1 INTRODUCTION

In Chapter 6 the importance of the matrix in a composite body was discussed. Resins form by far the largest group of matrix materials as far as carbon fibres are concerned, and hence they are discussed in some detail in this chapter. In addition, a brief survey of the most important work on metal matrices is given.

Resins can be divided into a number of groups according to their chemical formulation, but for use in composite manufacture it is proposed to divide them into three categories, namely:

1. Conventional thermosetting resins which can operate in the temperature range 125°–200°C.
2. High-temperature thermosetting resins which can operate at temperatures over 250°C, extending up to say 350°C.
3. Thermoplastic resins which by definition are limited to temperatures near ambient.

It is proposed to consider categories 1 and 2 in much more detail than 3 as most of the work on the production of high-strength and high-stiffness composites has involved resins of these types.

## 8.2    CONVENTIONAL THERMOSETTING RESINS

At the present time, the most important group of resins for use with carbon fibre are the epoxides, and for this reason they will be considered more fully than others.

### 8.2.1    EPOXY-RESIN SYSTEMS

Epoxy resins can be described as being highly versatile, as having a broad capability for blending with different hardeners, catalysts, etc., and, in addition, as being able to accommodate a wide variety of fillers and additives. Epoxy formulations can be soft, flexible, or hard, and they are available as solids or liquids. Epoxies are resistant to many environmental conditions and can be used up to 135°C on a continuous basis. A notable characteristic is their excellent degree of adhesion, which, coupled with relative ease of application, high strength, and good reproducibility, has led to their widespread use with carbon fibre.

Epoxies first found commercial acceptance in the period 1947–1950 and were utilised in coatings, in plastics tooling, and for adhesives. During this period epoxy resins started to be used in electrical insulation, while in the 1950s they found widespread application for many forms of protective coatings, where their properties of adhesion, toughness, and chemical resistance proved outstanding. It is hardly surprising, therefore, that there is a steady and continuous growth of the market for epoxy-based coatings and adhesives and that recent years have seen the development of advanced composite materials employing epoxy-resin matrices with glass reinforcement for use in aircraft and aerospace, where a high degree of success has been achieved.

Several epoxy resins currently being used in carbon fibre composites are listed below, and their properties are given in *Tables 8.1–8.3*.

*Shell Epikote 828/NMA/BDMA\**. This system normally consists of the following constituents, by weight:

| | |
|---|---|
| Epikote 828 resin | 100 parts |
| Epicure NMA hardener | 85 parts |
| BDMA accelerator | 1 part |

It is usual to mix an ample quantity of Epikote 828 and Epicure NMA, for which a mechanical stirrer is convenient. Without the addition of BDMA, the mixture will keep in a usable condition for one month at 20°C or longer at lower temperatures. One part of

* Benzyl dimethylamine.

**Table 8.1** PROPERTIES OF CAST EPOXY-RESIN SYSTEMS

| Property | ERLA 2772/ mPDA† | EPON 828*/ 1031/NMA‡ | ERLA 4305/ mPDA | ERLA 4617/ mPDA |
|---|---|---|---|---|
| Compressive modulus (lbf/in$^2$) | 441000 | 551000 | 987000 | 890000 |
| Compressive strength (lbf/in$^2$) | 19200 | 21600 | 35650 | 32800 |
| Tensile modulus (lbf/in$^2$) | 458000 | 507000 | 873000 | 783000 |
| Tensile strength (lbf/in$^2$) | 12900 | 9100 | 17300 | 19200 |
| Flexural modulus (lbf/in$^2$) | 462000 | 597000 | 912000 | 815000 |
| Flexural strength (lbf/in$^2$) | 17500 | 16400 | 30300 | 31000 |
| Heat-distortion temperature (°C) | 158 | 143 | 116 | 175 |

* Epon 828 is generally known as Epikote 828 in the UK.
† mPDA = m-phenylenediamene.
‡ NMA = nadic methyl anhydride.

BDMA is then added to 185 parts of the stock mixture just before use. The pot-life of the fully prepared resin is about 6 h. It should be noted that, when the BDMA is added, it needs to be well mixed to ensure uniformity, particularly since it is much lighter in weight than the resin or hardener and therefore tends to float on top of the mix. Furthermore, BDMA has a lower viscosity than the other two components.

The cure schedule is:

$\frac{1}{2}$ h at 150°C (curing operation)
2 h at 180°C (post-curing operation)

Higher post-curing temperatures increase the resin heat-distortion temperature and hence the useful working range of the composite.

*Union Carbide ERLA 2772/mPDA.* The cure schedule is:

2 h at 85°C (curing operation)
4 h at 160°C (post-curing operation)

*Shell Epikote 828/1031/NMA.* The cure schedule is:

2 h at 93°C and 1 h at 121°C (curing operation)
12 h at 150°C (post-curing operation)

*Union Carbide ERLA 4305/mPDA.* The cure schedule is:

4 h at 85°C and 3 h at 120°C (curing operation)
16 h at 160°C (post-curing operation)

**Table 8.2** PROPERTIES OF CAST ERLA 4617 SYSTEMS CURED WITH VARYING PROPORTIONS* OF MDA† AND MPDA

| Property | MDA 100% mPDA 0% | MDA 80% mPDA 20% | MDA 50% mPDA 50% | MDA 30% mPDA 70% | MDA 0% mPDA 100% |
|---|---|---|---|---|---|
| Heat-distortion temperature (°C) | 171 | 170 | 167 | 169 | 176 |
| Tensile strength (lbf/in$^2$) | 17 200 | 17 500 | 17 250 | 17 700 | 17 600 |
| Tensile modulus (lbf/in$^2$) | 651 000 | 643 000 | 710 000 | 724 000 | 782 000 |
| Elongation (%) | 5·1 | 5·1 | 3·4 | 3·5 | 2·9 |
| Flexural strength (lbf/in$^2$) | 25 300 | 26 000 | 28 200 | 28 500 | 31 400 |
| Flexural modulus (lbf/in$^2$) | 675 000 | 693 000 | 732 000 | 740 000 | 796 000 |
| Compressive strength (lbf/in$^2$) | 25 600 | 26 800 | 29 100 | 29 800 | 32 900 |
| Compressive modulus (lbf/in$^2$) | 687 000 | 701 000 | 742 000 | 789 000 | 848 000 |
| Deformation (%) | 9·1 | 9·1 | 8·6 | 7·8 | 8·1 |

* Blends in the resin system are based on 'amine equivalent weights'. All systems contain 110% stoichiometric addition of hardener.
† MDA = methylene dianiline.

**Table 8.3** PROPERTIES OF CAST ERLA 4617 SYSTEMS CURED WITH VARIOUS AMINE HARDENERS

| Property | mPDA | m-aminobenzylamine | p,p-diaminodiphenylmethane | ZZL 0820 | mPDA 70% ZZL 0820 30% |
|---|---|---|---|---|---|
| Heat-distortion temperature (°C) | 170 | 152 | 162 | 80 | 128 |
| Tensile modulus (lbf/in²) | 820000 | 700000 | 610000 | 550000 | 660000 |
| Tensile strength (lbf/in²) | 18500 | 17900 | 18800 | 14000 | 17000 |
| Elongation (%) | 2–8 | 3–2 | 6·0 | >12·5 | 4·0 |
| Flexural modulus (lbf/in²) | 860000 | 780000 | 690000 | 580000 | 730000 |
| Flexural strength (lbf/in²) | 31500 | 29500 | 30000 | 17500 | 29500 |
| Compressive modulus (lbf/in²) | 890000 | 850000 | 750000 | 600000 | 840000 |
| Compressive strength (lbf/in²) | 31500 | 30000 | 27000 | 17000 | 27000 |
| Deformation (%) | 7·5 | 7·0 | 8·0 | — | 6·0 |

Note that the properties of resins vary appreciably with the hardeners used.

*Union Carbide ERLA 4617/mPDA.* The cure schedule is:

4 h at 85°C and 3 h at 120°C (curing operation)
16 h at 160° C (post-curing operation)

Extensive information about the properties of epoxy-resin systems is also given by Soldats, Burhams, and Cole[1]. Typical properties of epoxy resins are summarised below for general guidance in using these materials.

*Tensile strength.* Based on the ASTM 638-52TD method, an unmodified DGEBA* resin cured with amine curing agents give values around 9000 lbf/in². With special curing agents, values approaching 20 000 lbf/in² may be achieved.

*Compressive strength and compressibility.* Based on the ASTM D695-54 method, typical results are 15000 lbf/in² for yield stress and 30000 lbf/in² for ultimate stress. It should be noted that the above test is only meaningful if it records the stress at which fracture occurs. If deformation continues without fracture, then the compressive yield figure is the more useful. If compressibility is due to squeezing-out of free volume, the limit is about 15% at a pressure of 12 000 atm.

*Flexural strength.* Based on the ASTM 790-49T method involving three-point bend tests, typical values are 18 000–25 000 lbf/in².

*Elastic modulus.* This can be determined from tension, compression, and flexural measurements based on the above ASTM methods, and the choice depends on the user. All three methods give similar values of the elastic modulus, which lies in the range $0.4$–$0.8 \times 10^6$ lbf/in², and is equal to about 1% of the modulus of carbon fibre.

*Poisson's ratio.* A typical value for a rigid epoxy system is $0.34$.

*Impact resistance.* Two basic types of test are in existence, namely the Izod type, carried out accordingly to the ASTM D256-54T method, and the falling-ball test, carried out according to the MIL-I-16923 ships' method. A typical value for the Izod test is $0.2$ ft lbf/in, and 2 ft lbf for the falling-ball test.

*Second-order transition temperature* $(T_g)$. This is a good measure of the working temperature range of a resin. A first-order transition

* Diglycidyl ether of bisphenol-A.

temperature is one at which a definite change in resin volume occurs as a result of applied temperature. However, a second-order transition temperature is one at which the rate of change of resin volume with applied temperature is not linear. Second-order transition temperatures must be higher than those relating to first-order transition. Below their $T_g$ temperature, polymers are rigid; above it, they are soft and elastomeric. The $T_g$ temperature is measured by hardness tests, bouncing-ball tests, dilatometer measurements, or heat-distortion temperature tests, according to individual preference in relation to end-use or application.

*Heat-distortion temperature.* This is typically 90°–180°C, depending on the curing agents used, but the employment of highly functional resins and curing agents increases this temperature to well over 200°C.

*Thermal expansion.* Below the $T_g$ temperature, a typical value is $39 \times 10^{-6}/$°C. Above the $T_g$ point, the value becomes larger and typical values can be as high as $100 \times 100^{-6}/$°C.

*Shrinkage.* This is of two kinds: curing shrinkage caused by the reaction during the curing operation, and thermal shrinkage due to cooling. It should be noted that most reaction shrinkage occurs while the resin is still in the liquid state. Accurate shrinkage values are difficult to obtain.

*Hardness.* A typical Rockwell M reading is 100.

*Thermal conductivity.* This is a difficult property to measure, but a typical value is $500 \times 10^{-6}$ cal cm$^{-1}$ s$^{-1}$ °C$^{-1}$.

*Specific gravity.* Typical values are 1·2–1·3.

### 8.2.2  OTHER THERMOSETTING RESINS

There are many types of thermosetting resins in addition to the epoxy systems, and some of the most important of these are mentioned briefly in this section. A glance at their most notable properties, methods of moulding, and typical applications will leave the reader in no doubt as to the versatility of the materials in this section. Moreover, their usefulness is growing day-by-day. Whilst each type was developed and has grown in usage quite outside the field of composites and the like, they must nevertheless be regarded as potential matrix materials. Use of resins from this group will allow the materials engineer to create composite structures having special properties to meet the needs of specific applications or will allow the best compromise to be achieved between properties, ease of fabrication, cost of manufacture, and so on.

*Phenolics.* As a class, these provide a combination of low cost, ease and versatility of moulding, and resistance to temperature, solvents, and chemicals unobtainable in other thermosets. Phenolics can be compression moulded, transfer moulded, injection moulded, or extruded. From the typical applications quoted below, it will be seen that this type of resin is frequently used as a matrix to bond various fillers which impart special characteristics. Mechanical properties are:

| | |
|---|---|
| Flexural strength | $10\,000$ lbf/in$^2$ |
| Maximum service temperature | Up to 200°C (mineral-filled) |

and applications are:

| | |
|---|---|
| Wood-filled | Handles, car instrument panels |
| Mica-filled | Computer plug boards, valve bases |
| Asbestos-filled | Heat-resistant lamp holders |
| Cotton-filled | Car dynamo fans, car fascia panels |

As stated in Part 1, phenolic resins have found widespread use in aerospace in ablative applications, for which considerable quantities of low-modulus carbon fibres have also been used. Phenolic resins are successful in ablative applications because they will char under the application of intense heat to yield a high carbon content. It is thought that, during the process of charring, the phenolic material undergoes a number of crosslinking reactions which ultimately yield a well-bonded and intact carbon structure. Failure to achieve crosslinking reactions would result in rapid disintegration of the carbon char. It is interesting to note that there is another class of ablative materials, namely those which sublime under the action of intense heat, and a number of polymers falls into this category.

Engelke, Pyron, and Pears[2] have reported the mechanical properties of Narmco 4028 phenolic carbon-fibre material and have given the thermal conductivity, heat capacity, and useful temperature ranges for phenolic carbon-fibre composite bodies.

*Alkyds.* These are thermosets and are superior to thermoplastics in resistance to creep and distortion at elevated temperatures and in abrasion resistance. However, they are less flexible and have lower impact resistance. An important characteristic is their excellent dielectric strength and low dielectric constant and power factor. These materials are available in four forms: putty or dough for encapsulation; granular; extruded rope, containing glass fibre; and in bulk for moulding.

Mechanical properties are:

| | |
|---|---|
| Tensile strength | 3000–4000 lbf/in$^2$ |
| Flexural strength | 7000–10000 lbf/in$^2$ |
| Tensile modulus | 1·5–1·7 × 10$^6$ lbf/in$^2$ |
| Flexural modulus | 2·0–2·5 × 10$^6$ lbf/in$^2$ |
| Impact strength | 0·11–0·15 ft lb/in |

*Amino compounds.* These are combinations of thermosetting, urea–formaldehyde, or melamine–formaldehyde resins with various fillers. They are usually compression moulded. Mouldings have excellent gloss with hard mar-resistant surfaces. Mechanical properties are dependent upon cure and water content. They exhibit linear stress–strain curves until fracture at about 1% strain. Applications are:

| | |
|---|---|
| Urea–formaldehyde | Domestic electrical accessories |
| Melamine–formaldehyde | Handles, high-performance electrical equipment |
| Mineral-filled melamine–formaldehyde | Heavy industrial electrical equipment needing high arc-track resistance and high heat resistance |

*Polyesters.* These are thermosetting resins generally used as matrix materials and usually available in the form of a transparent or translucent liquid. Composites can be made with them using contact moulding since heat and pressure are not required. Mechanical properties are:

| | |
|---|---|
| Tensile strength | 6000 lbf/in$^2$ |
| Flexural strength | 11000 lbf/in$^2$ |
| Flexural modulus | 0·4 × 10$^6$ lbf/in$^2$ |
| Barcol hardness | 45 |
| Shrinkage (liquid–solid) | 10% |
| Specific gravity | 1·2 |
| Water absorption at 20°C | 3% |
| Water absorption at 100°C | Crazes |

Polyester resins have found widespread applications as the matrix for glass fibre owing to their low cost and ease of fabrication.

F

## 8.3   HIGH-TEMPERATURE THERMOSETTING RESINS

Most of the high-temperature resins in use today have been developed during the last five or at most ten years. Much research effort has been devoted to the development of thermally stable polymers not only for composites for aircraft but also for many other applications such as high-temperature electrical insulation. Some of the materials so far developed have a service temperature close to 300°C, but ultimate goals are far higher, in the range 475°–500°C.

The following list of polymer systems is by no means complete but is none the less impressive and serves to indicate the scale of current research and development effort. It should be recognised that these materials are in many ways new or still under further development, which accounts for the lack of data on composites made from them, but much work on them is currently being carried out in materials laboratories throughout the world and especially in the USA. Some recent advances in this field were reported at the Ninth Electrical Insulation Conference, held in Chicago in September 1969[3]. It is important to note that high-temperature resins tend to be more difficult to use than conventional materials such as epoxies, which gives special significance to the actual fabrication techniques used if composites from different sources but containing the same fibre and resin are to have closely comparable properties.

### 8.3.1   AROMATIC POLYAMIDES

Aromatic polyamides are thermally very stable polymers, and Du Pont were probably the first company to introduce them, in 1961. They were sold commercially in the second half of the 1960s under the trade name Nomex. This type of polymer does not melt, is highly resistant to oxidation, and has very good electrical properties, which have led to its widespread use in the electrical field. It can be used continually at 220°C and for short periods up to 275°C, and hence it is classified as a high-temperature thermosetting resin although all the conventional polyamides belong to the thermoplastic group.

### 8.3.2   POLYAMIDE–IMIDES AND POLYESTER–IMIDES

This class of polymer is also very stable thermally. Polyamide imides with an application limit of 220°C are manufactured by Amoco Chemicals Corp. and are known as A1 Polymer. Westinghouse Electric Co. have also developed a similar polymer called

Aramidyl. Such materials can be used for high-temperature electrical insulation and for the manufacture of moulded components. The polyester–imides are available from the General Electric Co., who make Imidex-E resin (largely for wire coating), and Westinghouse, who make Enamel Omega for similar applications. Mobil Chemical Co. and P. D. George Co. are further examples of US manufacturers.

### 8.3.3 · POLYIMIDES

This type of polymer is recognised as having considerable potential application for high-temperature composites, in addition to normal moulding parts and coatings. In general, the high-temperature stability of polyimide resins is better than that of the previous two classes, and they are based on a combination of pyromellitic dianhydride and oxydianiline. In the USA, a number of manufacturers produces these materials, notably Narmco, TRW Inc., American Celanese, Du Pont, General Electric Co., US Polymeric, Monsanto, and others. The Narmco Division of Whittaker Corp. currently produces pre-preg tapes made out of glass fibre and a polyimide resin, Meltbond 840, the top application temperature of which is as high as 315°C. Narmco also make '1832', which is a B-stage polyimide designed for use in fibre pre-preg systems. TRW Inc. have a material known as P13N which is intended for use in composites and is claimed to be stable when combined with fibre in the pre-preg form; moreover, it is considered easy to use and its maximum service temperature can be up to 315°C. The General Electric Co. make Duromid resin for coating applications, and also Gemon 2010 and 3010 series reinforced moulding compounds, which can be used up to 250°C. A report in *Modern Plastics*[4] gives details of the General Electric Co. Gemon resins and indicates their potential suitability for high-performance jet engines. US Polymeric have a polyimide-based glass pre-preg called V-303, which is intended for the manufacture of composites by vacuum-bag moulding. Ferro Corp. is another US company active in the field of polyimide pre-preg systems.

In the UK Imperial Chemical Industries are active in the field of polyimide resins. A current ICI product is known as QX-13 and is reported to have short-term service at high temperatures extending up to 400°C.

Polyimide resins are generally tough, but they tend to be 'notch sensitive'. Moreover, they tend to have voids created in them during the curing operation. Typical mechanical properties are:

| Tensile strength | 12 000 lbf/in$^2$ |
|---|---|
| Specific gravity | 1·43–1·59 |
| Maximum service temperature | 260°C (to maintain good proportion of maximum strength) |
| Coefficient of thermal expansion | 15–17 $\times$ 10$^6$/°C |

Browning and Marshall[5] of the US Air Force Materials Laboratory describe work carried out on the production of carbon-fibre reinforced composites using polyimide resins. They worked with Monsanto's Skybond 700 and TRW P13N resins, using ozone-treated Thornel 40 and 50 fibre, and their composites exhibited flexural strengths in excess of 100 000 lbf/in$^2$ and short-beam interlaminar shear strengths in excess of 5000 lbf/in$^2$. These authors give a number of tables and graphs indicating that the loss in mechanical properties up to 315°C was very small. It is concluded that carbon fibre–polyimide composites can now be fabricated having a low void content with, say, 50% by volume fibre content and having mechanical properties comparable to those of epoxy-resin composites at room temperature but much better at elevated temperatures.

### 8.3.4   POLYBENZOTHIAZOLES

The Abex Corp. in the United States was one of the first to make this type of polymer commercially available. It appears to offer even better temperature stability than the polyimide type, making practical applications possible at 315°C. However, the curing procedures required are involved.

### 8.3.5   OTHER SYSTEMS

New systems are continually being developed, many of which depend on the polyimide structure. Other systems are centred around the polythiazole-type structure, while polyquinoxalines and their derivatives hold considerable promise in the never-ending search for more thermally stable resin matrices.

## 8.4   THERMOPLASTIC RESINS

### 8.4.1   GENERAL COMMENTS

A large number of thermoplastic resins exists today, and new compounds are continually being developed. The thermoplastic field has seen considerable growth, which is likely to continue at an accelerated pace. Materials of this type have found wide application, particularly in the chemical industry and for numerous products where their chemical inertness, toughness, and pleasing appearance make them invaluable. It is important to recognise that many of these resins have also been used successfully in combination with reinforcing materials such as glass or asbestos, in order to fulfil applications requiring greater strength or toughness.

Polymers such as polyethylene, polypropylene, and polyvinyl chloride have been established for many years, and a whole range of fabrication techniques is now available for them. They are relatively inexpensive and are particularly attractive where considerable weight of material is required, as in chemical-plant construction.

In recent years, there has been significant progress in the introduction of thermoplastic materials having higher operating temperature capabilities. Thus, materials of the fluorinated type, for example polytetrafluoroethylene, fluoroethylene propylene, and polyvinylidene fluoride, will maintain their shape and strength at higher temperatures than those mentioned in the previous paragraph; hence they may be classed as high-temperature thermoplastic materials. Indeed, one may regard the polyimides discussed in the previous section as being thermoplastic in some respects. However, when such resins are used in the production of high-temperature composites, it is perhaps logical to think of them as high-temperature thermosetting resins.

If one considers thermoplastic resins as a whole, it will be seen that most of the materials have a high viscosity at normal temperatures and can only be loaded with glass or carbon fibre, for example, with some difficulty. Work on the addition of carbon fibre to thermoplastics has been carried out on only a modest scale and with relatively little success; one of the major problems is the difficulty of loading carbon fibre, which is brittle and must be introduced as discrete bundles, as already described in Chapter 7. A further major problem is the low strength of the bond between the fibre and matrix interface. Since thermoplastics can exhibit considerable strain at relatively low stresses, they are not very compatible with carbon fibre. Thus, it is difficult for these materials effectively to transmit mechanical forces from one filament to another, and hence the overall properties of a thermoplastic composite tend to be very much lower than those based on a thermosetting matrix. Use of carbon fibre has nevertheless imparted some interesting properties to thermoplastic materials such as nylon or PTFE with regard to wear resistance and frictional characteristics, and these will be described in Chapter 10.

### 8.4.2   PROPERTIES OF THERMOPLASTIC RESINS

Brief details showing the properties of a few of the more widely used thermoplastic resins are given below, together with typical applications.

*Polyamides (Nylon 6, 6.6, 6.10, 6.11, 6.12).* These are thermoplastic and can be injection moulded. They can be machined and welded, also bonded by themselves. Mechanical properties are:

| | |
|---|---|
| Yield point | $\begin{cases} 11\ 000\ \text{lbf/in}^2\ \text{(Nylon 6)} \\ 10\ 000\ \text{lbf/in}^2\ \text{(Nylon 6.6)} \\ 8\ 000\ \text{lbf/in}^2\ \text{(Nylon 6.10)} \end{cases}$ |
| Elongation at break | 130% (Nylon 6) |
| Tensile modulus | $0.25 \times 10^6\ \text{lbf/in}^2$ (Nylon 6.10) |
| Melting point | $\begin{cases} 220°\text{C (Nylon 6)} \\ 286°\text{C (Nylon 6.6)} \\ 225°\text{C (Nylon 6.10)} \end{cases}$ |
| Maximum service temperature | $\begin{cases} 80°–100°\text{C (Nylon 6)} \\ 110°\text{C (Nylon 6.6)} \\ 65°\text{C (Nylon 6.10)} \end{cases}$ |
| Coefficient of thermal expansion | $\begin{cases} 7 \times 10^{-5}/°\text{C (Nylon 6)} \\ 9.9 \times 10^{-5}/°\text{C (Nylon 6.6)} \end{cases}$ |

Applications include bushes, bearings, handles, and wheels. Nylon 6.10 is used where dimensional stability is required, and Nylon 6.11 is used where high impact is required; in plasticised form it is almost unbreakable.

*PTFE.* These resins are high molecular weight polymers and do not melt. Processing is carried out while the material is in a gel state (at about 327°C). Moulding of PTFE requires a pre-form to be made which is sintered at 370°–380°C and cooled. Mechanical properties are:

| | |
|---|---|
| Tensile strength | $2000\ \text{lbf/in}^2$ |
| Maximum service temperature | 260°C continuous |
| Coefficient of thermal expansion | $9.9 \times 10^{-5}/°\text{C}$ |

Applications are dry bearing and piston rings.

*ABS (acrylonitrile–butadiene–styrene).* This can be injection moulded, extruded, or blow moulded. ABS resins fail in a tough ductile manner. Mechanical properties are:

| | |
|---|---|
| Yield stress | $7000\ \text{lbf/in}^2$ |
| Impact strength | 8 ft lbf/in |
| Water resistance | Very good |

Applications are safety helmets, car fascias, car body panels, and computer parts.

*Acrylics.* Apart from being handled in sheet form, these may be extrusion or injection moulded using higher than average pressures owing to high melt viscosity. Mechanical properties are:

| | |
|---|---|
| Yield stress | 6000 lbf/in$^2$ |
| Impact strength | $\begin{cases}4 \text{ ft lbf/in}^2 \text{ notched} \\ 19 \text{ ft lbf/in}^2 \text{ un-notched}\end{cases}$ |
| Maximum service temperature | 65°–95°C |

Applications include windscreens, machine guards, and meter cases.

*Acetal copolymers.* These may be processed by extrusion, injection moulding, or blow moulding. Mould shrinkage is 1·8–2·5%. Typical mechanical properties at 20°C are:

| | |
|---|---|
| Yield stress | 10 000 lbf/in$^2$ |
| Impact strength | 9 ft lbf/in$^2$ |
| | (notch tip radius = 0·080 in) |
| Maximum service temperature | 105°C |

Applications include car instrument housings, taps, cocks, and pulsators for washing machines.

*Acetal homopolymers.* These may be processed by extrusion at 200°C–215°C or by injection moulding at 200°C–220°C. Mould shrinkage is 1·9–2·4%. Mechanical properties are:

| | |
|---|---|
| Yield stress | 10 000 lbf/in$^2$ |
| Impact strength | 1·4 ft lbf/in |
| Maximum service temperature | 70°–75°C |

Applications are bushes, gears and bearings, pump parts, and valves.

### 8.4.3   REINFORCEMENT OF THERMOPLASTICS BY CARBON FIBRE

Thermoplastic materials on the whole do not wet carbon fibre at all, and hence it is extremely difficult to use unidirectional tows as the plastic matrix will fail to penetrate the latter. Consequently, most of the effort has been centred upon chopped fibre which may be intimately pre-mixed with the plastic matrix. Fibre chopped to various lengths has been used, but serious breakage or attrition invariably occurs during the mixing process and the final fibre length is usually much less than the starting length. The fibre is not aligned and is often poorly distributed; moreover, it is difficult to obtain high volume fractions of fibre owing to the high viscosity of most thermoplastic materials and, in practice, 15–20% represents the highest values. Use of sized fibre tows may be an advantage during the mixing as the tows are stiffer and less liable to fracture.

A representative collection of thermoplastic property data is given by Hollingsworth and Sims[6] and is shown in *Table 8.4*. This shows quite clearly that thermoplastics reinforced with chopped

**Table 8.4** PROPERTIES OF THERMOPLASTIC MATERIALS REINFORCED BY CHOPPED CARBON FIBRES*. From Hollingsworth and Sims[6], courtesy *Composites* and the authors

| Thermoplastic | Fibre content (% by volume) | Tensile strength (lbf/in$^2$) | Flexural strength (lbf/in$^2$) | Flexural modulus (10$^6$ lbf/in$^2$) |
|---|---|---|---|---|
| Polypropylene | 0 | 4400† | — | 0·21 |
| | 4·4 | 4900† | 7600 | 0·47 |
| | 8·9 | 5000† | 8080 | 0·66 |
| Toughened polystyrene | 0 | 4600 | — | 0·34 |
| | 4·9 | 5600 | 9950 | 0·77 |
| | 9·5 | 4150 | 9400 | 0·93 |
| Diakon | 0 | 9000 | 12900 | 0·46 |
| | 5·6 | 7900 | 14300 | 0·67 |
| | 10·7 | 7900 | 15200 | 1·10 |
| ABS | 0 | 6300† | — | 0·33 |
| | 5·0 | 8000 | 13700 | 0·68 |
| | 9·6 | 6100 | 10600 | 0·84 |
| Styrene–acrylonitrile | 0 | 10000 | 15400 | 0·48 |
| | 5·1 | 9500 | 14700 | 0·93 |
| | 9·9 | 9500 | 15150 | 1·50 |
| Phenoxy | 0 | 7050† | 11250 | 0·32 |
| | 5·5 | 7200† | 11600 | 0·58 |
| | 10·6 | 6300† | 11400 | 0·90 |
| Nylon 6 | 0 | 10000 | — | 0·34 |
| | 5·4 | 11250 | 19100 | 0·71 |
| | 10·2 | 13330 | 22900 | 1·12 |

* The fibre used was Type I, untreated. It is probable that a slight improvement would have been obtained with surface-treated material.
† Yield stress.

carbon fibre do not exhibit a significant increase in tensile or flexural strength, but their flexural modulus on the other hand is considerably increased. Moreover, the composites possess much greater dimensional stability, particularly at elevated temperatures.

At the time of writing it must be concluded that current fabrication techniques leave much to be desired and that significant advances in carbon-fibre reinforced thermoplastics will only come about by improved wetting and bonding to the fibre surface.

## 8.5 TERMINOLOGY

The steady growth of composite technology using a range of resins coupled with glass, asbestos, boron, or carbon-fibre reinforcement has led to the establishment of a number of special terms which are widely used in the manufacture of composites. A list of the more important terms relating to the qualities and properties of resins is given below.

### 8.5.1 SHELF-LIFE

The shelf-life is the period of time over which the flow and tack characteristics of a pre-preg sheet or tape are preserved when stored at a particular temperature. Ideally, a manufacturer would like one-component resin systems which were stable at room temperature. In practice, however, a shelf-life is limited to 3–6 months at a temperature of 5°C. Frequently, pre-preg materials are stored at − 18°C to ensure adequate shelf-life.

### 8.5.2 SHOP-LIFE OR WORKING LIFE

This is the period after the pre-preg has been removed from refrigeration over which the flow and tack characteristics are maintained. In practice, a 3 day period is the minimum acceptable, and 5–10 days is desirable and often attained. The longer the shop-life, the greater the flexibility for the user. Working life may be extended by leaving some solvent in the pre-preg. This also makes the latter more flexible. However, the solvent has to be removed during the final curing operation, which may as a result be slowed down. In addition, the presence of solvents at the final curing stage may give rise to internal voids and flaws.

### 8.5.3 CURE CYCLE

The cure cycle is the time–temperature profile selected for achieving optimum cure of a component. The cycle is constrained in practice by the tooling and by the ovens or autoclaves or presses used for the fabrication of the component. The resin selected must tolerate wide variations in heating rates if it is to be acceptable for a number of fabrication techniques. Heating rates of 1–5°C/min are representative of most autoclave-processed components.

### 8.5.4   STEPPED CURE CYCLE

A stepped cure cycle is a heating process where the rate of temperature rise is reduced to zero and a particular temperature is maintained for some selected time period. This kind of cycle is often used to ensure uniformity of temperature in both the component and the tooling, and to assist in obtaining a maximum flow of resin during the curing operation. The temperature is often selected to provide the longest period of flow and also the maximum amount of flow, which necessitates minimum resin viscosity before gellation of the resin occurs.

### 8.5.5   CURE TIME

The cure time is the amount of time required to effect a cure of the resin composition at the cure temperature. It is commonly a period of 1h, but shorter times are sometimes used for matched-metal mould processes, while in other cases longer periods are involved.

### 8.5.6   CURE TEMPERATURE

The cure temperature is the temperature required to effect polymerisation of the matrix or resinous component in the cure time.

Time and temperature are important variables in determining the rate of polymerisation of a resin matrix. However, for any given resin composition, a definable minimum cure temperature is that which is necessary to ensure at least 99% completion of reaction.

### 8.5.7   FLOW

Flow is a measure of the fraction of resin that exudes from the component being manufactured for the prescribed time, temperature, and pressure conditions employed in the process. This is a very complex property highly influenced by factors such as: (a) thickness of component; (b) resin content of the pre-preg; (c) heat-up rate; (d) time, (e) temperature; (f) pressure condition of B-stage of the pre-preg; (g) nature of catalyst or curing agent; and (h) volatile content of resin, if any.

A number of control experiments is often employed to establish flow and other properties which must be maintained during production. Procedures for determining flow require the establishment of set conditions for factors (a)–(h) above and the measurement of flow in terms of the ratio of resin exuded to the total weight of pre-preg (including fibre).

### 8.5.8   GEL TIME

Gel time is the time required for a given resin, held at a given temperature, to convert from a viscous fluid stage to a rubbery condition. It is intended to be a measure of the reactivity of a resin or the extent of the B-stage cure of a given resin. This property is used for production control purposes and to give the manufacturer an indication of the required processing conditions.

### 8.5.9   B-STAGE CURE

This is a partial cure of the resin. It may be defined as a relatively stable state in which the resin and hardener form an adduct which is brittle at low temperatures but will flow at high temperatures before gelation occurs. A resin after the B-stage cure is still flexible but is no longer a highly mobile liquid. Where solvents are used in the resin to improve flow characteristics, B-stage curing will result in the major part, but not all, being removed. After this stage has been reached, resins still have a definite shelf-life.

### 8.5.10   TACK

Tack can be defined as the sticking of a material: (a) to itself, (b) to the mould surface, or (c) to a polyfilm or release paper (as used in pre-preg manufacture).

A highly complex set of physical and mechanical phenomena is involved in tack, such as the amount of resin available to contact the surface, the extent of contact, the time allowed for contact, the temperature and pressure used in pressing the interfaces together, the time for which pressure and temperature are maintained, the roughness factors of the surfaces, and what is on the surface. Highly subjective, tack usually means having a material stick when it is supposed to and release when needed—part of the tailoring and services offered by resin compounders and pre-preggers. Pre-pregs must have the ability to adhere to themselves at ambient conditions under hand rubbing pressure. They should not be so sticky as to preclude removal of one ply from another if rearrangement or orientation of a layer of pre-preg becomes necessary. Pre-preg should be sufficiently tacky to adhere to a vertical mould surface after the surface has been treated with a release agent. Generally, however, after a mould has been coated with release agent, a resin gel coat is applied and the pre-preg is pressed against this, the gel coating helping to ensure a good surface finish.

### 8.5.11 VOLATILE CONTENT

Volatile content is a measure of the weight loss of a sample of pre-preg when exposed to a fixed temperature for a fixed time at a fixed air velocity. Intended to define the amount of residual solvent, it is often a measure of the volatile constituents of the resinous phase. When used for production control purposes, the volatile content is taken to be the ratio of the loss of weight at the B-stage to the original B-stage weight including fibre.

### 8.5.12 RESIN CONTENT BY WEIGHT

This is the percent by weight of resin taken on a volatile-free basis. Selection of the resin systems for carbon fibre may also incorporate consideration of functional properties of the cure system. Of major importance are heat-distortion temperature, oxidation resistance, and mechanical features of the resin such as its intrinsic tensile strength, its moduli (both shear and tensile) as well as its viscoelastic properties. It is often desirable to have resins of low modulus which have the ability to distribute the shear stresses.

## 8.6 METAL MATRICES

### 8.6.1 ADVANTAGES OF METALS

Until now, no mention of metals has been made. This is because of the concentration of effort so far on resin systems. Metals have only recently started to receive serious attention, and there is little doubt that interesting composite properties will be forthcoming from metal matrix systems. A noteworthy advantage of metals is their higher temperature capability when compared with even the most advanced resins. The main advantages of a metal compared with a resin matrix can be summarised as follows:

1. The specific strength of a metal is greater than for a resin, hence greater composite strength should be possible if in-built strains are not excessive.
2. Metals have a high tensile modulus, which is compatible with

carbon-fibre characteristics. Carbon-fibre reinforcement of metals can lead to a decrease in ductility and an increase in yield strength.

3. Metals possess very good high-temperature strength relative to resins, hence composites made from them should have a higher temperature capability.
4. Metals have a relative ease of forming and can be machined to close tolerances.
5. Metals have relatively good resistance to corrosion, but this depends to a large extent on the environment.
6. Metals have good resistance to impact.

### 8.6.2 PRODUCTION OF METAL-BASED COMPOSITES

In the production of a fibre–metal composite, the same general considerations apply as already discussed. Certainly, fibre alignment and ability to achieve a high volume percentage of fibre are important. Equally, it is necessary to ensure good adhesion between the surface of the fibre and the metal, just as with resin matrices, but, since the forming of dense fibre–metal composites often involves elevated temperatures, it is important to prevent a chemical reaction taking place between fibre and metal which would result in the formation of carbide interfaces.

The three most important methods of producing fibre–metal composites are:

1. Solidification of the metal melt which contains the fibre, preferably in the appropriate degree of alignment.
2. Sintering and pressing of a mixture of finely divided metal powder which contains the fibres.
3. Chemical or electrochemical coating of the metal on to the fibres. Electrodeposition is frequently used. The coated fibre may be further consolidated into a composite by means of sintering and pressing.

### 8.6.3 REVIEW OF RECENT WORK

Rolls-Royce Ltd have carried out work on the use of chemical decomposition techniques for coating and bonding metals to carbon fibres[7]. They report the use of tri-isobutyl aluminium, which could be chemically decomposed to give a metallic coating to the fibre, and the technique may be extended to nickel, cobalt, boron, and titanium. The work of Donovan and Watson-Adams[8] describes the use of electrodeposited nickel in carbon-fibre composites and shows the types of structure obtained together with fracture data. Additional work is being undertaken by universities and research establishments in the UK and the USA. Work with nickel is of particular significance in that nickel alloys are traditionally used for all kinds of high-temperature applications; it is not surprising, therefore, to find that one long-term goal of carbon fibre–nickel composites is high-temperature blades for aircraft engines, as distinct from carbon–resin blades which have been used at relatively low temperatures at the front end of air compressors.

Other metals such as aluminium may have applications when combined with carbon fibre. One potential use is for electricity transmission lines, the aluminium carrying the bulk of the current while the fibres provide high tensile strength allowing greater distances between transmission towers.

To date, very little information is available on these kinds of carbon-fibre composite. Moreover, it is expected that at least two years will elapse before reliable manufacturing techniques and composite data become established. Nevertheless, some initial work has been carried out by Jackson and Marjoram at Rolls-Royce[9, 10] on the compatibility of single carbon fibres coated with nickel or cobalt, and the results so far obtained indicate possibilities for the production of bulk composites. Rolls-Royce noted that coatings of copper and platinum are capable of weakening graphite fibre apparently without chemical reaction or structural recrystallisation taking place in the fibre tows, indicating the need for more development. An additional technique used by this company is electrolytically to deposit metals on the surface of the fibre followed by a rolling process in order to consolidate the coating and improve its properties. A general review of the state of the technology was given in 1965 by Cratchley[11] of Rolls-Royce. A more recent review extending to 1969 has been made by Weeton[12] of NASA Lewis Research Center.

In applying carbon-fibre reinforcement to metals, it is important to note that the relative stiffness of the fibre and the metal matrix is important and determines the final characteristics of the composite. For instance, Maire[13] of Le Carbone-Lorraine has indicated that a number of metals may be satisfactorily reinforced with carbon fibre, the fibre content extending up to 30% by volume. He found that, for temperatures over 350°C, the tensile strength of some composites increased, indicating greater ductility in the metal matrix and the ability of the matrix to transmit the load from fibre to fibre. Maire pointed out that fibres with a relatively low tensile modulus gave composites the modulus of which increased from 450°C upwards; however, when high-modulus fibres were used, the com-

posites gave modulus values which increased from room temperature upwards. The explanation for this behaviour lies in the relative
stiffness of fibre and matrix and also in their relative thermal
expansions. Thus, with low-modulus reinforcement, very little
contribution to composite stiffness is made at low temperatures,
but, at higher temperatures, the loss of stiffness by the metal matrix
is compensated for by the fibre. High-modulus reinforcement, on
the other hand, gives an improvement to the composite even at
room temperature.

Other factors are also involved, and studies of this kind illustrate
the complexity of reinforced metals and their behaviour but confirm
that improved properties at elevated temperatures can and have
been obtained.


## 8.7   OTHER MATRIX MATERIALS

### 8.7.1   RUBBERS

The amount of work carried out with rubbers is small. This is not
surprising as the characteristics of rubber compounds differ markedly
from those of carbon fibres. The main problem is to secure adequate
bonding between the fibre surface and the rubber matrix; failure to
achieve this will allow the latter to stretch freely and separate from
the fibre. Nevertheless, some degree of enhancement in properties
may occur. Thus, Genin[14] reports that carbon fibres may be used
to reinforce rubbers to give improved temperature resistance.
Carbon fibres may also be used in elastomers such as fluorocarbon
copolymers, and Bergstrom[15] reports improvements in the properties of these materials as a result of carbon-fibre additions. Sieron[16]
states that fluoroelastomer compounds reinforced with carbon
fibres exhibited a tensile strength retention in excess of 30% at
205°C, while the same compounds filled with carbon black retained
only 15% of the strength. Sieron[17] also reports work on butyl
rubber, ethylene–propylene terpolymer, and fluorelastomers with
fibre reinforcement. He found that dispersed (chopped) carbon fibre
was capable of increasing the strength of fluoroelastomers. For
practical application, however, carbon fibre must be cost effective
and, to achieve this, a significant improvement in performance is
needed relative to properties given by carbon blacks for example.


### 8.7.2   CERAMICS

As with rubbers, very little work has been carried out with ceramics and glass matrices. A major difficulty lies in fabrication and in matching of the thermal expansion characteristics of fibre and matrix. There is considerable incentive for more work in this area, however, owing to the high-temperature properties of ceramic materials, which will have wider application if increased strength and toughness can be imparted. Crivelli-Visconti and Cooper[18] of the National Physical Laboratory (NPL) have carried out work on the reinforcement of a pure silica matrix with up to 50% by volume of aligned carbon fibres. The reinforced silica composite exhibited a tensile modulus of $51 \times 10^6$ lbf/in$^2$ and an ultimate tensile strength of 230 000 lbf/in$^2$, and its porosity was in the range 10–20%. Crivelli-Visconti and Cooper carried out their mechanical property measurements based on a three-point bending test at room temperature. They found, in general, that weakly bonded composites showed high strength and toughness, while improved bonding decreased the toughness since cracks could propagate more easily.

### 8.7.3    CARBON

Much interest is being shown, particularly in the USA, on the fabrication of carbon fibre–carbon composites. Use of a carbon matrix allows very high temperature applications without loss of strength, and materials of this type are likely to find application where conventional materials have failed. Low-strength and low-modulus carbon fibres have been used for a number of years for reinforcing carbon, and one or two commercial products are currently available but their strength is comparatively low. Recently, success has been achieved in which high-modulus PAN-based carbon fibre (Modmor Type I) has been used to make a new type of carbon fibre–carbon composite. The development was carried out by McLoughlin at the Materials Engineering Laboratory of the General Electric Co.[19], who showed that composites of this type may have flexural strengths of as much as 140 000 lbf/in$^2$ at room temperature. Moreover, the composites retained their full strength at 1500°C in a helium atmosphere used to protect them from oxidation. McLoughlin believes that materials of this type have a better performance at such high temperatures than any other known substances.

REFERENCES

1 SOLDATS, A. C., BURHAMS, A. S., and COLE, L. F., 'Novel Cyclo-Aliphatic High Performance Epoxy Resins for Reinforced Structures', *Proc. 23rd Ann. Tech. Conf., SPI Reinforced Plastics/Composites Division,* Society of the Plastics Industry, New York (1968)
2 ENGELKE, W. T., PYRON, C. M., and PEARS, C. D., *Thermal and Mechanical Properties*

*of a Non-Degraded and Thermally Degraded Phenolic–Carbon Composite*, NASA Contractor Report NASA-CR-896, National Aeronautics and Space Administration, Washington (1967)

3  BURNS, R. L., 'Advances in High Temperature Polymers', *9th Electrical Insulation Conf., Chicago, 1969*

4  GENERAL ELECTRIC CO., 'New Polyamide for Easy Moulding', *Mod. Plast.*, **47** No. 3, 55 (1970)

5  BROWNING, C. E., and MARSHALL, J. A., 'Graphite Fiber Reinforced Polyimide Composites', *J. compos. Mater.*, **4**, 390 (1970)

6  HOLLINGSWORTH, B. L., and SIMS, D., 'New Fibre-Filled Thermoplastics: 2, Reinforcement by High Modulus Fibres', *Composites*, **1** No. 1, 80 (1969)

7  ROLLS-ROYCE LTD., Germ. Pat. 1 939 339 (19.2.70)

8  DONOVAN. P. D., and WATSON-ADAMS, B. R., 'Formation of Composite Materials by Electrodeposition', *Metals & Mater.*, 3 No. 11, 443 (1969)

9  JACKSON, P. W., and MARJORAM, J. R., Compatibility Studies of Carbon Fibres with Nickel and Cobalt', *J. Mater. Sci.*, **5** No. 1, 9 (1970)

10 JACKSON, P. W., 'Some Studies of the Compatibility of Graphite and Other Fibres with Metal Matrices', *Metals Engng Q.*, **9**, 22 (1969)

11 CRATCHLEY, D., 'Experimental Aspects of Fibre Reinforced Metals', *Metall. Rev.*, **10** No. 37, 79 (1965)

12 WEETON, J. W., 'Fibre Metal Matrix Composites', *Mach. Des.* **41** No. 4, 142 (1969)

13 MAIRE, J., 'Reinforcement of Light Metals by Carbon Fibres', Preprint of Paper 7.16, *3rd Conf. Industrial Carbons and Graphite, 1970* (to be published by Society of Chemical Industry)

14 GENIN, G., 'Pyrolysed Fibres for the Reinforcement of Plastics and Elastomers', *Plastiq. Infs,* **15** No. 318, 1 (1964)

15 BERGSTROM, E. W., *Improvement of High Temperature Properties, Part 2*, Report AD-609003, Rock Island Arsenal Laboratory (1964)

16 SIERON, J. K., 'Pyrolysed Fibers for High Temperature Reinforcement', *Rubb. Wld, N.Y.,* **148** No. 6, 50 (1963)

17 SIERON. J. K., 'High Temperature Elastometers for Extreme Aerospace Environments', *Rubb. Chem. Technol.*, **39** No. 4, 1141 (1966)

18 CRIVELLI-VISCONTI, I., and COOPER, G. A., 'Mechanical Properties of a New Carbon Fibre Material', *Nature, Lond.,* **221** No. 5182, 754 (1959)

19 MCLOUGHLIN, J. R., 'New High Temperature Carbon Fibre Composites', *Nature, Lond.,* **227** No. 5259, 701 (1970)

# 9

# Mechanical property
# measurements on
# carbon-fibre composites

## 9.1   INTRODUCTION

Methods of testing carbon fibres and their properties are described
in detail in Chapter 5. This chapter describes methods employed to
evaluate the performance of a composite. The testing of composites
is a relatively new field, and improved techniques are continually
being evolved; in any event, the scope of the subject is wider than
single-filament tests since there are many more measurements
which can be made.

The measurements described below* are based on the techniques
used by Morganite Modmor Ltd[1] for the making and testing of
composite specimens and involve: (a) tensile strength and modulus;
(b) compressive strength and modulus; (c) flexural strength and
modulus; and (d) shear strength.

Apart from impregnated-tow tensile tests, which always use
Araldite AY 103 resin, the tests described below are carried out on
specimens made from Epikote 828 resin. The latter was chosen as
it allows a well-tried and reproducible method to be used for
specimen preparation, to the benefit of anyone wishing to repeat

* The descriptions and figures given in Sections 9.2 and 9.3 are reproduced from the
Technical Data Sheet by Blakelock and Blasdale[1], courtesy Morganite Research &
Development Ltd.

the work. Composite testing, if carried out in depth, must involve a number of resin systems, and optimum conditions need to be established with any particular matrix.

## 9.2   PREPARATION OF SPECIMENS

### 9.2.1   PREPARATION OF COMPOSITE BARS FOR MECHANICAL TESTS

The method given here is only one of many possible alternatives, and it is the method used for routine checks on fibre properties. If the purpose of the tests is to characterise a particular carbon–resin system different from that used here, the method must be modified to suit the new resin. The necessary modifications are usually small for alternative resin systems that are suitable for wet-laying, but pre-preg resin systems may require critical appraisal. The technique followed when using pre-preg resins is to work from pre-preg sticks rather than pre-preg sheet since these generally give the best results obtainable with these systems. The results for pre-preg sticks may then be compared with those for pre-preg sheet to show the best form (i.e. thickness, volume fraction of fibre) of sheet to use.

The following procedure is used:

1. Epikote 828 resin is mixed as already described in Chapter 8.
2. The mould to be used is prepared by coating all its faces with release agents and warming to 50°–60°C. Release agents in general use are Mold-Wiz and Releasil 14. Both open-end and closed-end moulds are used in this work, but it is probably unwise to use open-end moulds for specimens having a thickness greater than 0·1 in.
3. If $A$ in$^2$ is the cross-sectional area of the composite to be made, $840A$ lengths of 10 000 filament tow are cut to form a bundle of fibre that should weigh $20A$ g/in. For a closed-end mould, the length of the cut tows should be equal to the length of the mould, whereas for an open-end mould the tows should be cut 2 in longer than the mould.
4. The bundle of fibres is placed in a suitable vessel, which may be a trough roughly made from thin aluminium foil. Resin is poured all over the fibres which are left to soak for 15 min.
5. Fibres are taken from the batch of resin and wiped of excess resin by gently pulling the bundle through the fingers (using rubber gloves).
6. The impregnated fibre is put into the warm mould, spacers are inserted to fix the thickness of the specimen, and the lid is put

in place and pressed gently down on the fibres by means of either clamps or a press.

7. The pressure on the mould is maintained and the specimen is gelled by treatment for 2 h at 100°C.
8. The specimen is cured at either 125°C for 24 h or 180°C for 2 h. This is usually done after the specimen has been removed from the mould, which can be carried out most easily whilst the mould is still hot.

The above method is used to prepare four types of standard specimens:

1. A 0·5 in × 0·1 in section in a closed- or open-end mould; approximately 42 tows in bundle, used for flexural and shear strength tests.
2. A 0·5 in × 0·25 in section in a closed-end mould; approximately 105 tows in bundle, used for impact strength tests.
3. A 0·5 in × 0·5 in section in a closed-end mould; approximately 210 tows in bundle, used for composite tensile and compressive tests.
4. A 1·0 in × 0·1 in section in a closed-end mould; approximately 84 tows in bundle, used for shear modulus measurements.

### 9.2.2   THE PREPARATION OF IMPREGNATED-TOW SPECIMENS FOR USE IN TENSILE TESTS

It is not general practice at Morganite Modmor to use correctly formed, 60% fibre fraction, composite specimens for routine tensile tests. Such specimens are costly and difficult to use, and they are therefore reserved for special purposes. The standard carbon fibre–resin tensile specimen is more accurately termed an impregnated tow.

The types of impregnated-tow specimens that have been used are shown in Figure 9.1. Figure 9.1(a) shows the older type of specimen which has reinforced ends and which was made in the kind of mould shown in Figure 9.2(a). This has now been superseded by the specimen shown in Figure 9·1(b) made in the rubber mould shown in Figure 9.2(b).

The new type of specimen is prepared as follows:

1. The mould is coated with release agent.
2. The density of the fibre to be tested is measured by floatation in a mixture of bromoform and carbon tetrachloride. For Modmor carbon fibre it is sufficiently accurate to assume values of 1·95 g/cm$^3$ and 1·70 g/cm$^3$ for Type I and Type II materials respectively.

3. One 6 in length of 10 000 filament tow is cut and weighed. The result for the mass per unit length of tow is recorded.
4. The resin is prepared by adding 6 parts by weight of Araldite AY 103 to 1 part of hardener. It is found that 15 g of the prepared resin will make six specimens.
5. The tow is painted with resin and placed in the mould.
6. Resin is poured around the fibres in the mould until all spaces are filled, and the ends of the two are splayed out into the flared parts of the mould.
7. The specimen is cured for 1 h at 100°C.

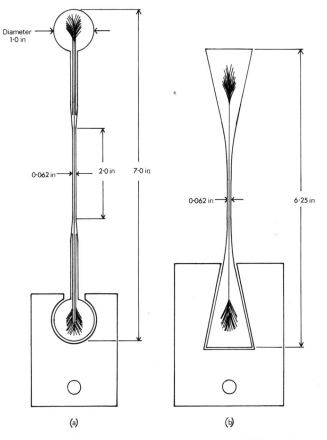

*Figure 9.1. Impregnated-tow tensile specimens (of uniform 0·062 in thickness): (a) old type; (b) new type*

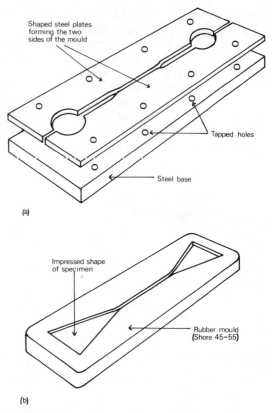

*Figure 9.2. Moulds for impregnated-two specimens: (a) old type; (b) new type*

## 9.3    METHODS OF TESTING

### 9.3.1    TENSILE MEASUREMENTS ON IMPREGNATED-TOW SPECIMENS

The specimens are mounted between special holders, shown in Figure 9.1, in an Instron machine fitted with a 1 in gauge length extensometer at the centre and pulled at a rate of 0·005 cm/min. The extension of the extensometer per pound-force of applied load is measured up to about half the expected breaking load (i.e. to about 90 lbf), and the extensometer is then removed. The test is completed by measuring the breaking load for the specimen.

Six specimens are used to characterise each batch of material. The results are calculated from the following equations:

$$\text{Type I fibre} \quad S_T = \frac{1{\cdot}89P}{M}$$

$$\text{Type II fibre} \; S_T = \frac{1{\cdot}60P}{M}$$

$$\text{Type I fibre} \quad E_T = \frac{1{\cdot}89}{Me}$$

$$\text{Type II fibre} \; E_T = \frac{1{\cdot}60}{Me}$$

where $S_T$ lbf/in² is the tensile strength of the fibre, $E_T$ lbf/in² is the tensile modulus of the fibre, $P$ lbf is the breaking load, $M$ lb/in is the mass of tow per unit length, and $e$ lbf⁻¹ is the extension of the specimen per unit length per unit load.

The numerical factor in these equations is equal to

$$\frac{\rho}{1 + [E_r(100 - V_f)/E_f V_f]}$$

where $\rho$ lb/in³ is the density of the fibre, $V_f$ is the volume fraction of the fibre (approx. 18%), $E_r$ lbf/in² is the tensile modulus of the resin, and $E_f$ lbf/in² is the tensile modulus of the fibre.

It should be emphasised that the four equations give values for the effective tensile strength and modulus of the impregnated fibre and not the properties of the specimen.

### 9.3.2   TENSILE TESTING OF COMPOSITES WITH HIGH FIBRE CONTENTS

The testing of resin-impregnated single tows is undertaken to measure fibre strength and not composite properties. Composite tensile properties can be measured on specimens machined from composite bars containing 50% or more fibre by volume.

A suitable form of specimen designed by the RAE, Farnborough, is shown in Figure 9.3. A composite bar of cross-section 0·25 in × 0·065 in and length 6·7 in is pressed. The central portion is then waisted using a 4 in diameter cutting wheel (preferably a diamond wheel) to a thickness of 0·040 in. Aluminium alloy plates of 16 s.w.g. are glued to the ends of the specimen with a cold-cure epoxy resin. For testing, the specimen can be held in wedge-action grips in an Instron machine. Such a specimen containing Type I fibre and having interlaminar shear strength not less than 2500 lbf/in² can be expected to break in tension, but shear failures near the aluminium

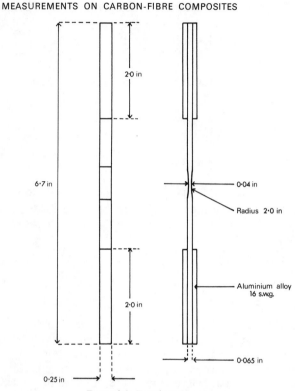

*Figure 9.3. Tensile test specimen*

alloy end-pieces may occur if the thickness of the specimen is greater than 0·065 in.

### 9.3.3  COMPRESSIVE MEASUREMENTS

Measurements of the compressive strength and modulus of composites are made using the specimens shown in Figures 9·4(a) and 9·4(b) respectively. The modulus specimen has the form of a uniform cylinder, 0·375 in in diameter and 1·75 in long. This specimen is therefore sufficiently long to be fitted with a 1 in extensometer for the measurement of the strain, but too long in relation to its diameter for accurate strength determinations. Modulus measurements are usually made up to about half the expected compressive breaking load.

The form of the compressive strength specimen now in use is shown in Figure 9.4(a). This specimen was also designed by the RAE, Farnborough, and is believed to be the most satisfactory specimen for tests with pure compressive stress. The central part of the specimen is reduced in cross-section to give a region of uniform stress, and the ends are fixed with resin into steel end-caps which have been machined to fit closely. When the resin used to fix the caps has been cured, the flat outer surface of each cap is machined

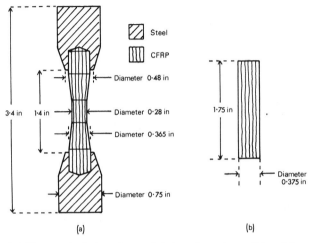

Figure 9.4. Compressive test specimens: (a) strength specimen; (b) modulus specimen

so that these are normal to the specimen axis and parallel to each other. The purpose of the caps is to prevent premature failure of the specimen due to splitting at the ends.

### 9.3.4 FLEXURAL MEASUREMENTS

Flexural strength and modulus measurements are made with composite bars 3 in long, 0·5 in wide, and 0·1 in thick. The specimens are strained in a three-point bend rig having a support span of 2·5 in. The span–thickness ratio in the tests is therefore 25. The Instron cross-head speed used in the measurements is 0·05 cm/min.

The flexural strength $S_F$ and the flexural modulus $E_F$ are calculated from the following formulae:

$$S_F = \frac{3PL}{2a^2b}$$

$$E_F = \frac{PL^3}{4a^3bd}$$

where $P$ lbf is the breaking load, $L$ in is the outside span, $a$ in is the thickness of the specimen, $b$ in is the width of the specimen, and $d$ in/lbf is the deflection at the centre of the specimen per unit load.

The results are quoted as the mean flexural strength and the mean flexural modulus for three specimens.

### 9.3.5   THE MEASUREMENT OF INTERLAMINAR SHEAR STRENGTH (HORIZONTAL SHEAR STRENGTH)

The interlaminar shear strength of composite bars is measured using bars $0.625$ in long, $0.5$ in wide, and $0.1$ in thick. In this case, the specimens are strained in three-point bending over a span of $0.57$ in.

The shear strength $S_S$ is calculated from the formula

$$S_S = \frac{3P}{4ab}$$

where $P$ lbf is the breaking load, $a$ in is the thickness of the specimen, and $b$ in is the width of the specimen.

Six shear-test specimens are broken for each result quoted. The load $P_T$ that will cause tensile failure in the shear specimens is given by the expression

$$P_T = \frac{2S_F a^2 b}{3L}$$

whereas the load for shear failure is

$$P_S = \frac{4S_S ab}{3}$$

Clearly, a specimen will fail in tension if $P_T$ is less than $P_S$ and in shear if $P_T$ is greater than $P_S$. For the specimens described here, $a = 0.1$ in and $L = 0.57$ in. Hence

$$\frac{P_T}{P_S} = \frac{aS_F}{2LS_S} = \frac{0.88\,S_F}{S_S}$$

Specimens will therefore fail only in shear if the shear strength is less than about $9\%$ of the flexural strength. Higher shear strengths will allow the specimens to fail in tension, and the measurement can then give no true value for the shear strength. In such cases, the value is quoted as 'greater than $x$', where $x$ is the value obtained by putting the load for tensile failure into the shear strength equation.

### 9.3.6 CONCLUSIONS

The most important properties referred to in this section are the tensile properties and the interlaminar shear strength. These have the closest relationship to the fibre properties and are therefore the most useful for process control. The flexural and compressive properties are too dependent on the nature of the resin matrix for this purpose and are therefore mainly useful for the characterisation of different carbon fibre–resin systems, which are not dealt with here. However, experience in the use of these methods has been obtained, and they are now available for the optimisation of composite preparation techniques and of curing schedules for fresh resins as these become of interest.

## 9.4 USE OF NOL RINGS FOR COMPOSITE EVALUATION

### 9.4.1 BACKGROUND

During the late 1950s and early 1960s, when reinforced plastics began to be widely tested, a large number of different techniques was used, leading to difficulties in correlating results. Consequently, when the US Naval Ordnance Laboratory (NOL) started to evaluate the effects of various glass-fibre finishes, they looked for a test which was quick, cheap, and accurate, paying particular attention to the case where the reinforcement consisted of non-woven glass filaments.

It was found that the most consistent composites were made in the form of rings by filament winding. These rings could be cheaply and easily made using the following dimensions:

| | |
|---|---|
| Internal diameter | 5·75 in |
| Width | 0·250 in |
| Thickness | 0·125 in |

These NOL rings may be tested whole in either tension or compression, or cut into segments for flexural and short-beam shear tests. The NOL ring test specimen has become well established and represents a proven test method. Although originally intended for glass fibre, it may be used for carbon fibre and is of particular significance for checking the quality of filament-wound articles.

### 9.4.2 FABRICATION OF NOL RINGS

NOL rings are made on a winder which has essentially three parts:

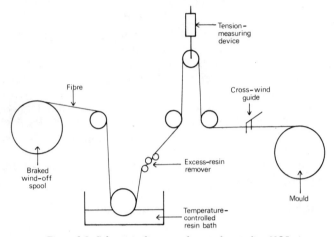

*Figure 9.5. Schematic diagram of a unit for winding NOL rings*

1. A fibre tensioning device, which may take the form of pulleys fitted with a braking device or Kinetrols.
2. A resin impregnation bath which is kept at a constant temperature while excess resin is removed from the fibre using a system of glass rods.
3. A mechanism for placing the fibre uniformly in a mould.

A schematic diagram of a typical winder is shown in Figure 9.5. The speed of rotation of the mould is variable, a typical speed being 6 rev/min. When the mould is full, the number of turns used is noted, the fibre is cut, and a band is placed around the outside of the mould. The mould is then transferred to an oven to cure the resin.

It is important to note that the fabrication must be carried out with great care if good reproducibility of results is to be achieved. As stated earlier in this chapter, all test specimens require careful preparation, and the NOL ring is no exception.

### 9.4.3   TESTING NOL RINGS

A NOL ring can be tested as a whole specimen or cut into segments for flexural or short-beam tests. The various mechanical properties are calculated according to the usual formulae:

$$S_F = \frac{3PL}{2a^2b}$$

$$E_F = \frac{PL^3}{4a^3bd}$$

$$S_S = \frac{3P}{4ab}$$

where $P$ lbf is the breaking load, $L$ in is the outside span, $a$ in is the thickness of the specimen, $b$ in is the width of the specimen, and $d$ in/lbf is the deflection at the centre of the specimen per unit load.

These formulae are subject to the normal errors involved in three-point bend tests and the additional error that the specimen is not flat.

Alternatively, the ring may be tested as a whole. A machine which puts the ring into compression does exist but is not in wide use. Most investigations break the ring in tension using split discs or hydraulic pressure. The split-disc set-up is shown schematically in Figure 9.6, and that for the hydraulic tests is shown in Figure 9.7.

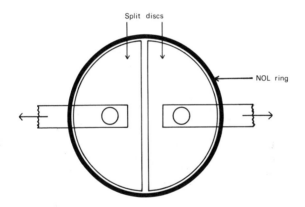

*Figure 9.6. Split-disc NOL-ring tester*

The split-disc method of test is very quick but has a number of limitations. The ring has its maximum tensile stress where the discs meet, and, since the load is not uniform, it is not possible to determine a modulus for the ring. While the ring is being loaded, it is also being deformed from its circular shape by a bending movement. This movement is at its maximum where the discs meet, hence that part of the ring is also subjected to a large bending stress. Thus, the strength measured is only an apparent tensile strength. Since the ring is forced to break at one or two points, the recorded strength

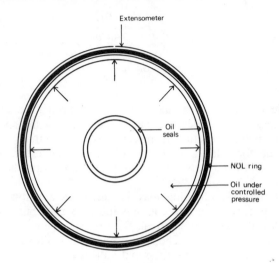

*Figure 9.7. Hydraulic NOL-ring tester*

may not be typical of the ring as a whole. The apparent tensile strength $S_T'$ of the ring when broken on split discs is given by

$$S_T' = \frac{P}{2tw}$$

where $P$ lbf is the break load, $t$ in is the ring thickness, and $w$ in is the ring width.

The hydraulic method of NOL-ring testing overcomes most of the above problems. The ring is hydraulically loaded and, since this must be uniform, no bending stresses are set up. Since the stress is uniform over the whole ring, a modulus measurement can be made by noting the increase in ring circumference for a given increase in hydraulic pressure. Uniform stress also implies that the ring will fail at its weakest points.

The tensile strength $S_T$ and modulus $E_T$ of the ring tested in this way are calculated from the following formulae:

$$S_T = \frac{p(\text{i.d.})}{2t}$$

$$E_T = \frac{p'(\text{i.d.})\,(\text{i.d.} + \text{o.d.})}{4c't}$$

where $p$ lbf/in² is the maximum hydraulic pressure sustained by the ring, i.d. and o.d. **denote** inside and outside diameters respectively (in inches), and $c'$ in is the change in ring circumference caused by a change of $p'$ lbf/in² in hydraulic pressure.

### 9.4.4   NOL RINGS CONTAINING CARBON FIBRE

There is no work reported by the SAMPE or SPI on NOL rings containing carbon fibre which have been broken hydraulically. Strength results have been obtained using split discs, and modulus measurements have been made by stretching complete rings between two pins and noting the diameter change for a given increase in load (typically 25 lbf). The latter is a very crude test method.

Davis and Zender[2] give the following values using Thornel 25 carbon fibre:
2256/ZZL 6820

| | |
|---|---|
| NOL-ring strength | $79\!\cdot\!4 \times 10^3$ lbf/in² |
| Fibre strength | $175 \times 10^3$ lbf/in² |
| Fibre volume fraction of ring | $0\!\cdot\!58$ |

Prosen[3] gives the following values using Hitco CY 1604 and ERLA 2256/ZZL 6820:

| | |
|---|---|
| Tensile strength | 70 600 lbf/in² |
| Tensile modulus (two-pin) | $3\!\cdot\!93 \times 10^6$ lbf/in² |
| Compressive strength | 60 500 lbf/in² |
| Compressive modulus | $3\!\cdot\!38 \times 10^6$ lbf/in² |
| Short-beam shear strength | 3 095 lbf/in² |
| Flexural strength | 80 500 lbf/in² |
| Specific gravity | $1\!\cdot\!37$ |

Desai and Kalnin[4] have also carried out experimental work and established a suitable procedure for filament-wound graphite-fibre reinforced composites. They fabricated several hundred NOL rings under different experimental conditions and tested them to determine their mechanical properties. Results indicated that, in general, there was effective translation of the individual fibre properties into the composite. These authors found that pre-heating of the graphite yarn was required to improve the interlaminar shear strength. As a result of establishing relations between all the filament winding process variables and the NOL-ring properties, they concluded that the highest strength and stiffness values were obtained from rings reinforced with surface-treated Type II Modmor fibre. *Table 9.1* summarises some of their experimental results.

**Table 9.1** FIBRE AND NOL-RING TEST DATA FOR VARIOUS FIBRES. After Desai and Kalnin[4]

| Fibre | Tensile modulus of fibre (10⁶ lbf/in²) | Tensile strength of fibre (lbf/in²) | Fibre content (% by volume) | Tensile modulus of ring (10⁶ lbf/in²) | Flexural modulus of ring (10⁶ lbf/in²) | Tensile strength of fibre (lbf/in²) | Flexural strength of fibre (lbf/in²) | Short-beam shear strength of ring (lbf/in²) |
|---|---|---|---|---|---|---|---|---|
| **WYB** glass fibre | 7 | 95 000 | 67 | 4·3 | 4·1 | 61 | 66 000 | 2 600 |
| **Thornel** 25 | 23 | 165 000 | 61 | 12·5 | — | 93 | — | — |
| **Thornel** 40 | 34 | 200 000 | 56 | 19·7 | 17·0 | 105 | 107 000 | 4 200 |
| **Modmor** Type I* | 57 | 195 000 | 42 | 20·5 | 16·6 | 74 | 108 000 | 8 100 |
| **Modmor** Type II* | 35 | 250 000 | 69 | 23·1 | 16·8 | 210 | 221 000 | 10 500 |

\* Surface-treated.

REFERENCES

1 BLAKELOCK, H. D., and BLASDALE, K. C. A., *Mechanical Property Measurements on Modmor Carbon Fibre–Resin Composites*, Morganite Research & Development Ltd, London (1968)
2 DAVIS, J. G., and ZENDER. G. W., 'Mechanical Behaviour of Carbon-Fibre Reinforced Epoxy Composites', *Advances in Structural Composites, Vol. 12*, Society of Aerospace Materials and Process Engineers/Western Periodicals Co., North Hollywood (1967)
3 PROSEN, S. P., 'Carbon Fibre NOL Ring Studies', *Proc. 21st Ann. Tech. Conf., SPI Reinforced Plastics Division*, Society of the Plastics Industry, New York (1966)
4 DESAI, R. R., and KALNIN, I. L., 'The Effect of Filament Winding Process Variables on the Performance of Carbon or Fibreglass Reinforced NOL Rings', *Proc. 24th Ann. Tech. Conf., SPI Reinforced Plastics/Composites Division*, Society of the Plastics Industry, New York (1968)

G

# 10
# Properties of carbon-fibre composites

## 10.1 INTRODUCTION

At the time of writing, a considerable amount of evaluation work is being carried out on the mechanical properties of carbon-fibre composites using test methods described in Chapter 9. The vast majority of this work is based on composites having epoxy or high-temperature resin matrix systems. Some typical composite property data are presented in this chapter for the more straightforward measurements of parameters such as the ultimate tensile strength and the modulus in tension, flexure, and shear. It must be recognised, however, that additional characteristics such as fatigue properties, toughness, properties at elevated temperatures, and corrosion resistance are also important in the design of components.

A point often overlooked when quoting test results is the need to specify the test methods used; this allows components made and tested in one location to be compared directly with those made in another.

## 10.2 SHEAR-STRENGTH PROPERTIES

In the early work on fabricating carbon-fibre composites, a major problem arose owing to the low interlaminar shear strength (ILSS) values obtained. They occurred as a result of poor bonding between the fibre surface and the resin matrix. The importance of this bond has already been discussed in Chapter 6. Although it has been shown that the presence of fibre reinforcement in more than one direction can overcome this deficiency to some extent, good bonding is nevertheless essential to obtain optimum composite properties.

Considering composites made from the early batches of PAN-based carbon fibre, it will be seen that they gave the following typical ILSS values (based on epoxy resins and the short-beam method of testing already described in the previous chapter):

$$\text{High-modulus fibre}\quad 2500\text{--}3000\ \text{lbf/in}^2$$
$$\text{High-strength fibre}\quad 6000\text{--}7000\ \text{lbf/in}^2$$

These values are much lower than the $12\,000\text{--}20\,000\ \text{lbf/in}^2$ tensile strength of epoxy resins.

Simon, Prosen, and Duffy[1] carried out work at the US Naval Ordnance Laboratories on the shear strengths of carbon-fibre composites. They found that untreated fibre gave poor ILSS results but, if the fibre was treated with nitric acid prior to incorporation into the composite, shear strengths up to $6700\ \text{lbf/in}^2$ could be obtained with high-modulus fibre, presumably owing to a modification of the fibre surface giving improved bonding. There are indications that good shear strength is essential if the compressive strength of a composite is to be satisfactory; hence, there was good reason from the start of carbon-fibre composite technology to establish means of obtaining better bonding between fibre and resin. It is interesting to note that the problem is more pronounced with the high-modulus fibre since it has a graphitic-type structure; graphite is not readily wetted by most substances, and bonding is difficult to achieve.

Composites subjected to bending will break down as a result of accumulation of fibre fractures on the tension side of the test bar if the composite shear strength is high enough. Simon and his co-workers have suggested ILSS values as high as $15\,000\ \text{lbf/in}^2$ as a target level to ensure that this happens or, in other words, to ensure that the carbon fibres are effectively utilised in the composite.

It is interesting to note that in the USA attempts were made to improve the shear-strength characteristics of carbon fibres and resins by the use of techniques such as 'whiskerising'. Much of this work was carried out by Prosen and Simon[2] in close co-operation with Milewski, Shaver, and Withers[3], who had developed a technique for growing silicon carbide whiskers on to rayon- or PAN-based carbon fibres. Shaver[4], of General Technologies Corp., describes the whiskerising technique whereby minute single crystals are grown perpendicularly from the carbon-fibre surface and serve to provide an improved bonding between fibre and resin. Shaver also describes the fabrication of composites from whiskerised fibres. Prosen and Simon report improvements in shear strength of graphite fibre–epoxy composites from approximately 3000 lbf/in$^2$ to over 11 000 lbf/in$^2$ and suggest that values as high as 18 000 lbf/in$^2$ may be reached. The use of this method never became widespread, however, owing to the high cost of growing the whiskers and to the increased difficulty of wetting-out the whiskerised fibre with resin. Nevertheless, the work helped to encourage the development of carbon-fibre composite materials.

Recent work carried out on PAN-based carbon fibre has shown that it is possible to surface treat the material in order to achieve improved ILSS values. This is referred to in Chapter 4. Thus, for treated fibre, the following results are typical of those obtained with epoxy resins and short-beam testing:

High-modulus fibre  Up to 11 000 lbf/in$^2$
High-strength fibre  Up to 15 000 lbf/in$^2$

Even more recently, values of ILSS as high as 18 000 lbf/in$^2$ have been obtained by the Narmco Materials Division of Whittaker Corp.[5] using PAN-based carbon fibre, together with their own 1004 resin system and special curing.

The surface-treatment process differs entirely from the whiskerising work: first, it does not deposit or coat the fibres in any way; secondly, it does not interfere with the ease of wetting the fibre surface with resin; and thirdly, the process is relatively cheap to carry out. As a result of these developments, surface-treated carbon fibre has become accepted as a necessity for good composite manufacture.

## 10.3   OTHER MECHANICAL PROPERTIES OF PAN-BASED CARBON-FIBRE COMPOSITES

### 10.3.1   PROPERTIES BASED ON EPOXY RESIN SYSTEMS

*Table 10.1* gives some typical US results for the properties of Modmor PAN-based fibre–epoxy resin composites[6] (these data were obtained at the US Air Force Materials Laboratory and the US Naval Ordnance Laboratory). It will be seen that some variation of composite properties does occur, particularly with respect to flexural strength. These variations may be partly explained by differences in the composite-making techniques used. Nevertheless,

**Table 10.1** PROPERTIES OF MODMOR CARBON FIBRE-EPOXY RESIN COMPOSITES

| Fibre type and form | Composite type | Fibre content (% by volume) | Composite specific gravity | Strength (lbf/in²) | | | | Modulus (10⁶·lbf/in²) | | | |
|---|---|---|---|---|---|---|---|---|---|---|---|
| | | | | Tensile | Flexural | Shear | Compressive | Tensile | Flexural | Shear | Compressive |
| Type I short | Unidirectional | Not reported | — | — | 73 900 | 2 900 (untreated) | — | — | 35·0 | — | — |
| | | | — | — | 110 000 | 5 500* | — | — | 41·8 | — | — |
| | | 65 | 1·73 | — | 124 900 | 7 000 | — | 32·3 | 28·5 | — | — |
| | | 65 | 1·73 | — | 122 900 | 7 500 | — | 39·9 | 26·4 | — | — |
| Type II short | | Not reported | — | — | 168 000 | 7 800† (untreated) | — | — | 22·7 | — | — |
| | | 70 | 1·61 | — | 220 000 | 8 000 (untreated) | — | 22·8 | 18·3 | 0·91 | — |
| | | 66 | — | — | — | — | 143 900 | — | — | — | 26 |
| Type II continuous | | 59 | 1·51 | 175 000 | 140 000–200 000 | 6 200 (untreated) | 126 000 | 20·6 | 20 | — | 17·3 |
| | | 65 | 1·58 | — | 218 200 | 7 900 (untreated) | — | — | 19·2 | 0·93 | — |
| Type I short | Bidirectional 8-ply | 60 | 1·71 | — | 77 700 | 3 700 | — | 17·9 | 17·2 | 0·93 | — |
| Type II short | | 60 | 1·55 | — | 128 700 | 4 100 | — | 11·1 | 12·1 | 0·81 | — |
| Type II continuous | | 65 | 1·55 | — | 116 500 | 3 700 | — | 10·5 | 13·9 | 0·84 | — |

* 6310 lbf/in² after 6 h boil in water.
† 5400 lbf/in² after 6 h boil in water.

a number of very high flexural strength and modulus values is shown, for both Type I and Type II fibres. The data also show that composites made with short fibre and with long fibre possess similar flexural strengths and moduli.

Work has been carried out by Dauksys and Ray[7] of the US Air Force on the fabrication and properties of composites made with carbon fibre from a number of sources. It is believed that the results obtained by them are among the most reliable so far available on epoxy-based composites, and these are shown in *Table 10.2*. The resin system used by Dauksys and Ray was Union Carbide ERLA 2256/ZZL 6820. From the table it will be seen that the PAN-based fibre, namely Courtauld's Grafil and Morganite Modmor, gave particularly good values for flexural strength, and modulus, compared with the rayon-based fibres. The authors noted that the value of compression strength obtained from the Modmor Type II composite was higher than any observed for a graphite-based fibre–epoxy matrix composite. It should be noted, however, that compression tests are difficult to perform and results tend to show appreciable variation.

At the time of writing, it is believed that composites made from PAN-based carbon fibres generally give superior properties compared with those from fibres based on other precursors. The reasons are not fully understood but are in part due to the good balance between the fibre tensile strength and modulus, which gives breaking strain values in the range 0.3–1.0%, coupled with the ability to form a good bond with the resin matrix.

### 10.3.2    PROPERTIES BASED ON HIGH-TEMPERATURE RESIN SYSTEMS

Recently, information has been published by ICI on the properties of carbon fibre used in conjunction with their QX-13 polyimide resin[8]. This resin was specifically developed for application with carbon fibre, glass fibre, and glass cloth. Details of the properties of QX-13, methods of use, and composite properties are given in the ICI literature, from which *Table 10.3* is taken (data in the table were supplied by the RAE, Farnborough).

It will be seen that properties given in *Table 10.3* refer only to Modmor Type I (high-modulus) fibre. The value for flexural strength is given as 120000 lbf/in$^2$ at room temperature, which compares closely with the value of 124900 lbf/in$^2$ obtained by Dauksys and Ray using an epoxy resin (see *Table 10.2*). However, the flexural modulus ($22 \times 10^6$ lbf/in$^2$) is low compared with the Dauksys and Ray value of $34.7 \times 10^6$ lbf/in$^2$. One possible explanation for the reduced modulus with QX-13 resin is that the fibre

**Table 10.2** AVERAGE LONGITUDINAL PROPERTIES OF UNIDIRECTIONAL GRAPHITE–EPOXY COMPOSITES. From Dauksys and Ray[7], courtesy *Journal of Composite Materials*

| Composite | Fibre content (% by volume) | Composite specific gravity | Strength (lbf/in²) | | | Modulus (10⁶ lbf/in²) | | |
|---|---|---|---|---|---|---|---|---|
| | | | Flexural | Short-beam shear | Compressive | Tensile | Flexural | Compressive |
| Thornel 25 | 62·1 | 1·36 | 117 000 | 5200 | 74 900* | 13·9 | 15·2 | 12·7* |
| Thornel 40 | 64·0 | 1·41 | 124 600 | 4700 | — | 19·2 | 17·2 | — |
| Thornel 50 | 65·7 | 1·48 | 114 900 | 3600 | 81 400* | 24·0* | 23·1 | 23·8* |
| Thornel 50S (treated) | 60·3 | 1·51 | 144 900 | 7200 | 88 000* | — | 22·4 | 23·0* |
| HMG 25 | 61·5 | 1·37 | 109 000 | 5500 | 63 600 | 10·6 | 8·8 | 9·6 |
| HMG 50 (E80-95 finish) | 65·7 | 1·52 | 153 700 | 6400 | 73 600 | 27·8 | 22·8 | 31·6 |
| Great Lakes (staple yarn) | 58·4 | 1·59 | 107 000 | 4300 | — | — | 25·0 | — |
| Great Lakes (tow) | 56·7 | 1·58 | 112 500 | 3700 | — | — | 21·7 | — |
| Modmor Type II (epoxy sized) | 64·4 | 1·58 | 218 200 | 7900 | 143 900* | 21·0* | 21·8 | 26·0* |
| Modmor Type I (treated, epoxy sized) | 62·6 | 1·74 | 124 900 | 7000 | 66 900* | 32·3 | 34·7 | 45·5 |
| Courtaulds HT† | 71·8 | 1·60 | 168 200 | 7100 | 74 900 | 29·3 | 25·3 | 26·8 |
| Courtaulds HM† | 67·8 | 1·65 | 173 000 | 8700 | 82 700 | 33·3 | 31·8 | 31·6 |
| Courtaulds pre-preg warp sheet | 60·6 | 1·55 | 193 800 | 7200 | 110 500 | — | 19·5 | 19·8 |
| Modmor Type II pre-preg, Narmco 5505 resin | 60·0 | 1·57 | 184 000 | 11 000 | — | — | 18·0 | — |

* Data obtained from different samples of approximately the same fibre content.
† Owing to pronounced data scatter, the properties of four panels were averaged and tabulated. The fibre content of the four HT composite panels ranged from 70·3% to 72·6% and that of the four HM composite panels from 66·9% to 69·6%.

**Table 10.3** TYPICAL PROPERTIES OF UNIDIRECTIONAL QX-13 LAMINATES WITH MODMOR TYPE I (TREATED) CARBON FIBRE (FIBRE CONTENT 52% BY VOLUME). Courtesy Imperial Chemical Industries Ltd, Plastics Division

| Property | Test temperature | Value |
|---|---|---|
| Specific gravity | | 1·67 |
| *Flexural strength\** | | |
| Before ageing | Room | 8 400 kgf/cm$^2$ / 120 000 lbf/in$^2$ |
| After ageing in air at 200°C for 25 h | Room | 8 400 kgf/cm$^2$ / 120 000 lbf/in$^2$ |
| After ageing in air at 200°C for 100 h | Room | 8 400 kgf/cm$^2$ / 120 000 lbf/in$^2$ |
| After ageing in air at 200°C for 1000 h | Room | 8 300 kgf/cm$^2$ / 120 000 lbf/in$^2$ |
| After ageing in air at 200°C for 1550 h | 200°C | 7 400 kgf/cm$^2$ / 100 000 lbf/in$^2$ |
| After ageing in air at 300°C for 25 h | Room | 7 700 kgf/cm$^2$ / 110 000 lbf/in$^2$ |
| After ageing in air at 300°C for 100 h | Room | 7 700 kgf/cm$^2$ / 110 000 lbf/in$^2$ |
| After ageing in air at 300°C for 1000 h | Room | 6 300 kgf/cm$^2$ / 90 000 lbf/in$^2$ |
| After ageing in air at 300°C for 1770 h | 300°C | 3 500 kgf/cm$^2$ / 50 000 lbf/in$^2$ |
| *Interlaminar shear strength†* | | |
| Before ageing | Room | 450 kgf/cm$^2$ / 6 400 lbf/in$^2$ |
| After ageing in air at 200°C for 1000 h | Room | 350 kgf/cm$^2$ / 5 000 lbf/in$^2$ |
| After ageing in air at 300°C for 1000 h | Room | 270 kgf/cm$^2$ / 3 900 lbf/in$^2$ |
| *Flexural modulus* | | |
| Before ageing | Room | 15 × 10$^5$ kgf/cm$^2$ / 22 × 10$^6$ lbf/in$^2$ |
| After ageing in air at 200°C for 1000 h | Room | 15 × 10$^5$ kgf/cm$^2$ / 22 × 10$^6$ lbf/in$^2$ |
| After ageing in air at 300°C for 1000 h | Room | 13 × 10$^5$ kgf/cm$^2$ / 18 × 10$^6$ lbf/in$^2$ |

\* Flexural strengths were measured using a span–depth ratio of 24·1 and a centre anvil of $\frac{1}{2}$ in radius.
† Interlaminar shear strength was measured by the short-beam three-point bending test using a span–depth ratio of 6:1.

volume is only 52%, compared with 62·6% with the epoxy resin.

Based on the limited work carried out with high-temperature polyimide resins, there are strong indications that composites can be produced from them having room temperature mechanical properties approaching those based on epoxy systems. However, at elevated temperatures up to, say, 300°C, the former are distinctly superior, the epoxides showing rapid loss of strength at temperatures exceeding 150°–175°C.

One problem in the fabrication of polyimide-based composites has been the tendency for voids and internal porosity to remain after curing. Use of improved techniques involving vacuum bags and autoclaves has helped to overcome this drawback. Moreover, the newer polyimide resins possess improved handling and fabrication characteristcs. Thus, the polyimide matrix has now become fairly well established owing to its good mechanical properties at room temperature coupled with excellent property retention at elevated temperatures.

### 10.3.3 FATIGUE PROPERTIES

Fatigue testing is receiving attention in both the UK and the USA. To date, however, relatively little information exists, but the general view is that carbon-fibre composites possess very good fatigue characteristics and appear to be superior to aluminium alloys on a weight basis. *Table 10.4* shows the results of some early tests on a blade of carbon fibre–resin composite.

**Table 10.4** RESULTS OF FATIGUE TESTS ON A COMPOSITE MADE FROM UNTREATED TYPE I FIBRE AND EPOXY RESIN

| Test | Tensile load (lbf) | Bending moment stress (lbf/in$^2$) | Oscillatory stress (lbf/in$^2$) | Cycles (10$^6$) |
|------|------|------|------|------|
| 1 | 0 | 0 | 10000 followed by 15000 | 100 100 |
| 2 | 0 | 0 | 10000 followed by 12000 | 100 145 |
| 3 | 0 | 0 | 17500 | 120 |
| 4 | 1260 | 15600 | 5200 | 50 |
| 5 | 1100 | 13200 | 4850 | 33 |

None of the specimens failed.

### 10.3.4    EFFECT OF WATER

Carbon fibres themselves are not affected by immersion in water. Where the resin matrix is also unaffected, the mechanical properties of carbon-fibre composites are therefore not significantly changed by prolonged immersion in water at ambient temperatures. With some resins, however, a slight loss in strength, up to 5%, may occur, but the stiffness will remain unchanged. Nevertheless, evidence has been obtained indicating that at elevated temperatures a small amount of water is absorbed by most resins, and this acts as a mechanical plasticiser thereby causing a loss in strength greater than 5%.

### 10.3.5    EFFECT OF TEMPERATURE

Carbon fibres retain their full mechanical strength up to at least 1000°C. With resin-based composites, the resin matrix is the strength-limiting factor. Thus, with a particular resin system, the data already presented in Chapter 8 will determine the effect of temperature on composite properties. It should be pointed out that carbon will commence to oxidise in air at 350°–400°C, and hence the matrix must provide adequate protection. Polyimide-type resins are not as effective as epoxies, for example, in excluding air; hence, the resin may be a limiting factor in preventing oxidation, but temperatures of 350°C and above are sufficiently high to cause loss of mechanical properties in even the most advanced resins anyway.

### 10.3.6    TEMPERATURE COEFFICIENT OF EXPANSION

The temperature coefficient of expansion of the fibres themselves is small and negative in the longitudinal direction. A test carried out on a carbon fibre–polyester composite gave a coefficient of $-0.73 \times 10^6/°C$ along the fibre direction, and $+29 \times 10^6/°C$ perpendicular to the fibre.

### 10.3.7    HARDNESS AND MACHINABILITY

The hardness of the majority of carbon fibre–resin composites is very close to that of the particular resin used as the matrix. Since most resins possess a relatively low hardness value compared with metals, little difficulty is experienced in carrying out machining operations. In practice, it is possible to machine carbon-fibre composites in exactly the same manner as one would machine the parent resin.

It follows that, for the successful machining of composites, tools must be sharp and it is preferable to machine the components in a dry condition. The most satisfactory results have been obtained using high speeds since this procedure improves the surface finish and reduces burning during the cutting operation. For practical purposes, the cutting speed is limited only by the tendency of the tool to overheat, thereby losing its cutting edge. For lathe work, cutting speeds of 200 ft/min are quite satisfactory using high-speed tools but, where tungsten carbide tools are employed, speeds of over 300 ft/min are preferable. It is difficult to make general comments on feed rates, but up to 0·010 in per revolution of the work piece is recommended, depending on the surface finish required. For heavy cutting, a positive top rake of about 20° is advantageous with side and front clearance of approximately 15°. For finishing cuts, no top rake is necessary.

Milling operations can be carried out just as easily as on plastics materials, but again care must be taken to avoid chipping which may arise if heavy cuts are taken. Where the highest quality of finish is required, it is necessary to follow conventional machining by fine grinding using conditions laid down by grinding-wheel makers for plastics materials.

### 10.3.8   SPECIFIC ELECTRICAL RESISTANCE

The specific electrical resistance of the high-modulus fibre along its axis lies in the range 750–800 $\mu\Omega$ cm at 25°C, depending on the graphitising temperature used. At higher temperatures, the resistance actually falls, in keeping with the well-known temperature–resistance relationship for bulk carbon. Thus, at 200°C the value is about 650 $\mu\Omega$ cm. In a carbon-fibre composite, the resistance along the fibre direction is proportional to the length of the fibres and inversely proportional to their total cross-sectional area. In practical terms, there is some difficulty in arriving at precise values since one has to make good electrical contact with all the fibres at each end of the composite. At right angles to the fibre direction, the resistance is naturally much greater as the current flow is impeded by the matrix phase and the results tend to be more variable. Clearly, cross-plies in composites even out the resistance characteristics in any direction.

## 10.4   FRICTION AND WEAR PROPERTIES OF PAN-BASED CARBON-FIBRE COMPOSITES

Carbon fibres exhibit some interesting friction and wear properties when combined with various resins. Much of the work undertaken in this area has been carried out by Lancaster and his colleagues at the RAE, Farnborough. Thus, Lancaster[9] reports experiments designed to study the effect of reinforcement by carbon fibres on the friction and wear of polyester resin sliding either against itself or against tool steel. The work of Lancaster and his colleagues showed that carbon-fibre reinforcement can reduce significantly both the coefficient of friction and the rate of wear. The maximum effect was achieved when the fibre axes were oriented normal to the sliding surface, and Lancaster suggested that the fibres act as preferential load-bearing regions of contact.

Further work carried out by Giltrow and Lancaster[10] in which a wide range of polymers was reinforced by carbon fibre shows that, whereas the coefficient of fraction of the polymers ranged from 0·25 to 0·75, the addition of 25–30% by weight of fibre gave a common level of from 0·25 to 0·35 for all the composites. Moreover, the authors found that the wear rates of the reinforced plastics were dramatically reduced compared with the straight plastics in all cases. Factors of 1000 and over were observed, indicating that the carbon fibres supported most of the load. Again, the best results were obtained with the fibres residing normal to the friction surface.

It must be pointed out that the majority of plastics dry bearings are made from PTFE or nylon and, as already seen in Chapter 8, these materials are difficult to load with carbon fibre, particularly in an aligned condition. It appears, therefore, that much additional work, particularly on the fabrication techniques for incorporating the fibre into the plastic matrix, will be necessary before the potential of carbon fibre can be fully utilised.

## 10.5   CHEMICAL PROPERTIES

Relatively little specific information is available on the chemical properties of carbon fibre–resin composites and data will need to be established by extended trials for different environments. It is possible, however, to obtain useful guidelines by considering the behaviour of the reinforcement and matrix components separately.

Carbon, and hence carbon fibres, are well known for their chemical inertness towards many chemicals and reagents. Much information is available in the literature relating to carbon in bulk form which should be useful as an indication of its behaviour in filament form. In particular, carbon fibre is much less prone to attack by strong alkalis than is glass fibre, but, on the other hand, strong oxidising agents are likely to be detrimental to it. Glass-fibre composites have found widespread use in areas where chemical inertness is required, the fibres being used in association with a number of resins. There does not appear to be any reason why carbon-fibre composites should not ultimately find many similar applications, particularly in the chemical industry.

In the chemical industry, polyethylene and polypropylene are now used extensively for applications demanding chemical inertness; fluorinated thermoplastics have found steadily increasing use in specific applications where particularly severe environments are encountered; PVC, nylon, and other thermoplastics have also established a range of uses and tend to show good chemical inertness. *Table 10.5* indicates the relative chemical inertness of four well-known resins towards a number of reagents.

**Table 10.5** INERTNESS OF RESINS TOWARDS A NUMBER OF CHEMICAL REAGENTS

| Chemical | Resin | | | |
|---|---|---|---|---|
| | *Epoxy* | *Polyester* | *Phenolic* | *Furane* |
| Water | No reaction | No reaction | No reaction | No reaction |
| Alkalis | Good | Fairly good | No good | Good |
| Inorganic salts | Good | Good | Good | Good |
| Dilute mineral acids | Good | Good | Good | Good |
| Strong mineral acids | No good | Fairly good | Fairly good | Fairly good |
| Fuel gas, petrol, and oils | Good | Good | Good | Good |
| Crude oils | Good | Good | Good | Good |
| Organic acids | Fairly good | Fairly good | Good | Good |
| Soaps and detergents | Good | Good | Fairly good | Good |
| Ketones, esters, and ethers | Fairly good | No good | Good | Good |

In considering chemical properties in general, it is important to recognise that the matrix material is the one which is primarily exposed to the particular environment. Consequently, it offers protection to the fibre which calls for good bonding between itself and the fibre surface. Where chemical plant is subjected to mechanical loading which imparts alternating strains in the material, the ability of the matrix and fibre to remain in intimate contact will influence the resistance to chemical attack. Thus, if the resin matrix is made to stretch and contract owing to the loading, any tendency for microcrazing will reduce the chemical inertness and may allow the fibre to be attacked. Where carbon fibre is used as reinforcement, the very low strain exhibited by the fibre is an advantage since it will minimise the amount of strain borne by the matrix and hence reduce microcrazing.

## 10.6    BEHAVIOUR OF CARBON-FIBRE COMPOSITES TOWARDS IMPACT AND FRACTURE

The property data given so far in this chapter refer to static or slow applications of loading on specimens and serve to indicate their performance relative to other bulk materials such as metals. Although much information is available on the behaviour of the bulk materials under impact loadings, comparatively little is currently known about the impact behaviour of carbon-fibre composites. It is not surprising, therefore, to find that much work is being undertaken in this area.

The studies of the fracture behaviour of PAN-based carbon-fibre composites carried out by Bader, Bailey, and Bell[11] are of particular interest. Their results indicate that the resistance to impact and fracture of such composites is strongly dependent on the fibre–matrix bond, the fibre strength, and fibre volume fraction. Bader and his colleagues present evidence to show that very good fibre–matrix bonding tends to reduce toughness, while the ability of a composite to delaminate under impact increases toughness as it allows the impact energy to be dispersed through the composite. They found that, the higher the fibre strength and the greater the strain to fracture, the larger the resistance to impact; thus, the high-strength form is better in this respect than the high-modulus form since it can accommodate suddenly applied stresses to a greater extent. In the case of PAN-based fibres, the high-strength form exhibits twice the strain at breaking load compared to the high-modulus type. Based on these results, the higher the volume fraction of fibre in the composite, whether high-strength or high-modulus, the greater its toughness.

Work on the fracture of composites has shown that, when they fail as a result of impact, a two-stage breakdown frequently occurs. In the first stage, compression buckling takes place on the impacted face and this progresses towards the neutral plane. In the second stage, tensile rupturing of the fibre occurs on the face opposite to that receiving the impact, thereby initiating tensile pull-out of the fibres. Figures 10.1(a) and 10.1(b) show respectively the first and second stages of the breakdown. As a result of this and other studies of fracture behaviour, it may be stated that the requirements for maximum toughness in composites are:

1. Good fibre–matrix bonding, but very high bond strengths to be avoided.
2. Maximum fibre strength.
3. High fibre strain.
4. Maximum fibre content.

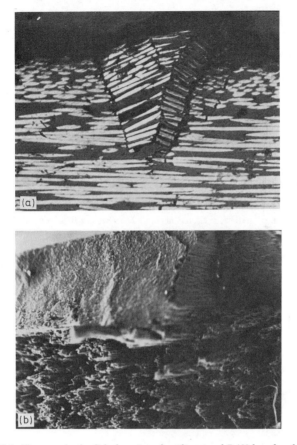

*Figure 10.1. Photograph of polished section of unidirectional PAN-based carbon-fibre composite: (a) the top face, showing the point of impact and buckling of the fibre; (b) the back face of the specimen away from the point of impact. Courtesy Bader, Bailey, and Bell*[11]

Much additional study will have to be undertaken in order to establish a full understanding of the mechanism of composite breakdown and, in particular, the role of fibre and matrix properties together with that of their bonding.

### REFERENCES

1 SIMON, R. A., PROSEN, S. P., and DUFFY, J., 'Carbon Fibre Composites', *Nature, Lond.,* **213** No. 5081, 1113 (1967)

2 PROSEN, S. P., and SIMON, R. A., 'Carbon Fibre Composites for Hydro and Aerospace', *Plast. Polym.,* **36** No. 123, 241 (1968)

3  MILEWSKI, J. V., SHAVER, R. G., and WITHERS, J. C., 'Whiskerized Graphite Filaments for Composites', *Mater. Engng,* **67** No. 5, 62 (1968)
4  SHAVER, R. G., 'Silicon Carbide Whiskerized Carbon Fibre', Pre-print 24E,
   *Materials Conf., Philadelphia, 1968,* American Institute of Chemical Engineers,
   New York
5  WHITTAKER CORP., private communication (1970)
6  BLAKELOCK, H. D., and LOVELL, D. R.. 'High Modulus Reinforcing Carbon', *Proc.
   24th Ann. Tech. Conf., SPI Reinforced Plastics/Composites Division,* Society of
   the Plastics Industry, New York (1969).
7  DAUKSYS, R. J., and RAY, J. D., 'Properties of Graphite Fibre Non-Metallic Matrix
   Composites'. *J. compos. Mater.,* **3** No. 4, 684 (1969)
8  IMPERIAL CHEMICAL INDUSTRIES LTD. *Polyimide Resins: Laminating Resins QX13.*
   (1970)
9  LANCASTER, J. K., 'Composite Self Lubricating Materials', *Proc. Instn mech.
   Engrs,* **182** No. 2, 33 (1967–68)
10 GILTROW. J. P.. and LANCASTER, J. K.. 'Friction and Wear Properties of Carbon
   Fibre Reinforced Materials.' *Wear.* **12** No. 2, 91 (1968)
11 BADER, M. G., BAILEY, J. E., and BELL, I., *The Fracture Behaviour of Carbon Fibre
   Reinforced Composites,* University of Surrey, Department of Metallurgy and
   Materials Technology, private communication (1970)

# Part 3

Part 3

# 11

# Applications for carbon-fibre composites

## 11.1 INTRODUCTION

High-strength and high-modulus carbon fibres are very costly at the
present time when compared with conventional materials of construction, but prices have fallen appreciably in the last two years.
High prices are mainly due to the product being at the development
or early production stage, but it must be appreciated that the
manufacturing process is fairly complex, involving specialised heat-
treatment operations. Nevertheless, it is certain that, as true production levels are reached, substantial reductions in manufacturing
costs will be achieved, this, in turn, allowing greater usage and still
further reduction in cost.

At the time of writing, almost all fibre production is being used in
the aircraft and aerospace industries, where its relatively high cost
can be justified by the saving in weight. These industries are able to
show that carbon-fibre composites are cost effective in selected
areas, which accounts for the high level of interest being shown. It is
worth noting that composites as such are not new to the aircraft
industry, which has steadily increased its usage of glass fibre–resin
components in order to save weight or reduce manufacturing costs.
Much additional experience has been gained in the USA with high-
strength and high-stiffness boron composites, and this has further
helped to pave the way for carbon fibres.

It is anticipated that the major outlet for high-performance carbon fibres will lie in the aircraft industry for the next two or three years at least. During this period, many of the engineering problems associated with these new materials will be sorted out, which will lead the way to wider application in other industries. This chapter considers the progress made in the aircraft and aerospace fields and indicates the other likely areas and industries which will benefit from carbon-fibre composites in due course, particularly when price levels become more competitive with established engineering materials.

## 11.2   AIRCRAFT AND AEROSPACE

### 11.2.1   AIRCRAFT STRUCTURES EXCLUDING ENGINES

At the present time, a number of aircraft components is being made out of carbon fibre–resin composites. The following list gives an idea of their variety:

| | |
|---|---|
| Ailerons | Helicopter tail booms |
| Air ducts | Landing-gear doors |
| Bulkheads | Leading edge panels |
| Fairings | Spoilers (air brakes) |
| Flaps | Stabilisers |
| Floor sections and floor beams | Storage vessels |
| Fuselage sections, spars, etc. | Tail assemblies |
| Fuselage skins | Wing tips |
| Helicopter blades and rotor shafts | |

Many of the above components are Class 1 structures, which implies that, if failure occurs, the safety of the aircraft is threatened. For this particular class, long and exhaustive testing under flying conditions is required before final approval can be given. Class 2 and Class 3 structures are not so critical, but, nevertheless, much testing is required. Included in Class 1 are all structural components and control surfaces together with certain internal fittings such as aircraft seats. Items such as coat racks, galleys, etc., are not vital to an aircraft's safety and have a lower class rating.

It is normal practice to make part of a wing tip or a single air brake, for example, out of the composite material, followed by test flying, if the initial engineering test data obtained on the ground is sufficiently promising. In this way, failure will not jeopardise the aircraft. More extensive usage follows success with the initial stage. Many hours of testing, both on the ground and in the air, are required to prove the new material, the whole operation taking up to two years per component. It is only after one or two components have been fully approved together with consistency data (variations within a batch) that they may be specified in current production programmes. The cost of such work is very great indeed, much of it being financed with the aid of military or government contracts.

A list of aircraft incorporating carbon-fibre and boron-fibre composites is given in *Table 11.1*. This list is not complete but indicates the range of activities at the time of writing.

A number of references is given at the end of this chapter indicating the nature and scope of work going on in the USA and in Europe. A *Financial Times* survey[1] gives general information on carbon-fibre applications with emphasis on its use in aircraft. Of particular interest are the wing tips made from a composite material incorporating Modmor fibre and a Narmco epoxy resin[2], which, it was stated, had satisfactorily completed over eight hours test flying on an A7 Corsair 2, an aircraft manufactured for the US Navy and Air Force by the Ling Temco Vought Corps. It is believed that this was the first PAN-based carbon-fibre composite component to be used successfully in flight, and it was 34% lighter than a glass–epoxy resin equivalent.

The investigations of McElhinney, Kitchenside, and Rowland[3], at the British Aircraft Corp. Ltd, Weybridge, are also interesting as they involved aluminium beam sections which were strengthened and stiffened by bonding CFRP* to them in appropriate areas. As a result, a weight saving was achieved with less development effort than required for an all-composite beam. BAC were particularly interested in weight saving on their proposed 3-11 aircraft and intended to employ CFRP where possible, obviously by stages. One problem with the attachment of carbon fibre to a metal structure is the differential expansion of the metal and composite; in the BAC programme, its effect was minimised by use of carefully controlled curing cycles during fabrication.

References 4–9 describe other work, in the USA, on the use of CFRP in aircraft components. Some of the work has been undertaken with US rayon-based fibre while other uses PAN-based material produced in the UK. Almost all the work described has been carried out with government support and involves military aircraft.

BAC are the first to have flown a carbon-fibre structural component in a British aircraft[10]—the BAC 167 Strikemaster, which was fitted with a rudder-trim tab made entirely from carbon fibre–resin

---

* CFRP denotes carbon-fibre reinforced plastic. This is an abbreviation for carbon fibre–resin composite. Glass fibre–resin composites are often known as GRP materials (see Chapter 1).

**Table 11.1** AIRCRAFT INCORPORATING CARBON FIBRE*

| Country | Manufacturer | Aircraft | Component |
|---|---|---|---|
| UK | British Aircraft Corp. | 3-11 | Airframe components, floor beams |
| | British Aircraft Corp. | 167 Strikemaster | Rudder-trim tab |
| | British Aircraft Corp. | Jet Provost | Trim tab |
| | Hawker-Siddeley | Harrier | Wing tip, antennae, access doors |
| | Hawker-Siddeley | 748 | Torque tube |
| | Westland | Wasp helicopter | Tail section, rotor spars, leading and trailing edges of rotor blades, doors, transmission shafts, floor beams |
| Sweden | Saab Scania | 37 Viggen | Access panels, wheel doors |
| USA | General Dynamics | F-111 | Wing pivot doubler (boron), wing roof fairing (carbon) |
| | Grumman | F-14 | Landing gear doors, access doors |
| | Ling Temco Vought | A7 Corsair 2 | Wing tips |
| | Lockheed | C5A | Floor panels |
| | North American Rockwell | F-100 | Wing section |
| | Northrop Norair | F-5 | Wing-tip leading edges |
| | Teledyne-Ryan | Firebee | Stabilator |

* This list is not complete but indicates the range of activities at the time of writing.

and aluminium honeycomb and developed under a Ministry of Technology contract. Morganite Modmor supplied the fibre, which was pre-pregged by Fothergill & Harvey Ltd, using Union Carbide ERLA 4617 resin.

Westland Helicopters Ltd[11, 12] is another British company active in the carbon-fibre field. They are investigating the possibilities of CFRP for blades, airframes, transmission units, flooring, and tail booms in helicopter construction.

Schwartz[13] reviewed the availability in mid-1969 of high-strength and high-modulus composites, together with their properties and prospects. Apart from the well-established glass–epoxy systems, boron fibres and carbon fibres are included in his review, according to which carbon fibres are largely used with epoxy resins and also with high-temperature polyimides. Mention is made of the Narmco NMD 1004 and Modmor Type II fibre, which are stated to show promise of doubling the best previous tensile and shear strength properties.

Almost all the above work on CFRP materials for use in the aircraft industry is at the research and development stage, requiring relatively small quantities of fibre. Nevertheless, a great deal of testing and evaluation has already been done. Based on the successes already achieved with small aircraft components, it has been established that carbon-fibre composites actually work and achieve weight saving. However, it must be stressed that at least two years will elapse before true production runs commence on any selected component, if only because of the need to carry out all the testing on a very thorough basis as air-worthiness certificates demand convincing proof from any new material. One factor is certain: the aircraft industry will demand high levels of consistency from the fibre in terms of mechanical properties, and its appearance/handling characteristics will also need to be of a high standard so that composites can be fabricated having truly consistent properties. There is no evidence today that aircraft makers are anxious to have a higher ultimate tensile strength or modulus in their reinforcement materials as the present properties appear to be quite adequate. However, they demand a high level of consistency. No clear view can be given at present on the relative merits of high-modulus or high-strength fibre, which depend on the component and its function, and it is likely that both types will be used.

It will be seen that the current evaluation work is concerned entirely with structural parts of aircraft, and this will ultimately lead to a fair proportion of their structure being made out of composites. In the area of air transport, however, further big possibilities exist. Today, there is a large and growing world network of air freight, involving the handling of cargo in large tonnage quantities. It is in this area that carbon fibres are likely to see sizable outlets—for the fabrication of pallets, containers, and tanks to hold the cargo, all of which contribute to the weight of aircraft but not the payload and where even a modest saving in weight can have a big influence on airline operating economies over extended periods.

## 11.2.2    AIRCRAFT ENGINES

Although the principle of the jet engine is simple, the problems facing the engineer in its development are great. Modern engines have a remarkable power output for their size and weight, made possible only by very advanced materials of construction. In particular, the bladed rotors operate at tremendous speeds, creating problems of blade construction. The centrifugal forces acting on the blades impart high stresses at the roots, and vibrations set up in the blades cause flutter. At the high-temperature end of the engine, special nickel alloys are used extensively; here the blades are relatively small and the major problem is one of high-temperature creep. At the front cold end of the engine, the blades are of larger size, particularly in the case of the by-pass engine, and they compress the air prior to entry into the combustion chamber and by-pass ducts. In these large blades, very high centrifugal loads are experienced owing to their size and rotational speed; moreover, variation of the air flow pattern and resonance in the blades can cause excessive flutter, and it is the normal practice to use titanium because of its high strength–weight and high stiffness–weight ratios. Titanium also possesses a high degree of toughness.

Carbon-fibre composites, having exceptionally high specific strength and stiffness, are particularly attractive for the front end-blade application. Since a CFRP blade may be lighter in weight than a titanium equivalent, the engine will be less heavy. Moreover, the lighter weight blades will experience a lower centrifugal force, allowing a weight saving at the blade roots and mounting which, in turn, will allow a lighter rotor disc. Additionally, the greater stiffness of the blades will allow them to operate with less flutter and hence greater efficiency.

Rolls-Royce have carried out the most notable work aimed at improving jet-engine efficiency by incorporating carbon fibre in all the appropriate components. To achieve their objectives, they have developed their own PAN-based carbon fibre and associated prepreg and composite technology. The prime goal has been the introduction of carbon-fibre blades on the large-diameter fan of the

RB-211 by-pass engine. This application is undoubtedly the most significant to date for carbon-fibre composites and could result in large quantities of fibre being used. Gresham[14] has indicated that a CFRP blade for the RB-211 engine weighs about 10 lb and at maximum speed exerts a centrifugal load on the disc of 36 tonf. If a titanium blade of the same design were used, its extra weight would require the disc to be $2\frac{1}{2}$ times as strong which, in turn, would entail a heavier shaft, bearings, and supports. There can be little doubt that the jet-engine blade application is one of the toughest and most severe for any material, metal, or composite, but the saving of around 300 lb per engine or, say, 1200 lb per four-engined aircraft by using CFRP is a target well worth striving for.

The manufacturing techniques evolved by Rolls-Royce centre on a continuous carbon-fibre pre-preg tape unit[15]. In this process, the PAN fibre is wound on to frames and stitched together so that the tows are parallel and in tape form prior to oxidation. Oxidation and further heat treatment follows, and the carbon fibre is then coated with resin to make a tape of the required thickness. The tape is known as Hyfil, a registered trade mark of Rolls-Royce. The blades are made by laying up successive layers of the pre-preg tape cut to the required shape and with the fibres in the appropriate direction. A patent taken out by Rolls-Royce[16] gives details of a method of laying up PAN-based fibres impregnated with epoxy resin. With this method, the blade can be built up to the required profile. The actual moulding of the $3\frac{1}{2}$ ft long blades has to be carried out with particular care, involving heat and pressure in a precision-steel mould. A swivelling-head power press has been specially built for Rolls-Royce by Power Moulding Ltd[17]. A feature of this press is that the two halves of the mould open in such a way that the top platen faces upwards and can therefore be thoroughly cleaned between each pressing. Clean moulds are essential to the successful fabrication of large precision components such as these. It is noteworthy that the mould size can be up to 6 ft × 5 ft with the above unit, and loads up to 200 tonf may be applied. Additional presses having special features for the moulding of composite structures have been produced by Bradley and Turton Ltd[18].

Detailed investigations have been carried out by Rolls-Royce to establish the optimum design for a blade and also to find the best means of attaching the blade to the hub or disc[18-23]. In general, the Hyfil blade has a higher resonant frequency both in torsion and in bending than the equivalent titanium blade, thereby reducing flutter to a marked degree. It is particularly noteworthy that the lay-up of the fibre controls the resonant frequency characteristics;

thus, it is possible to alter these without changing the shape of the blade if conditions in the engine demand it.

In order to speed up their development as much as possible, Hyfil components were made as replacements for existing Rolls-Royce Conway engines and, in collaboration with BOAC, 20 engines were converted. The first engine so modified ran in September 1967 and was granted a certificate of air-worthiness in April 1968. In May 1968, VC-10 aircraft on African routes started to fly with three normal engines and one with carbon-fibre blades, thereby allowing considerable flying experience to be obtained. However, difficulties have been experienced during these VC-10 trials owing to

*Figure 11.1. Photograph of an early development RB-211 engine showing the fan blades made of carbon-fibre composite materials. Courtesy Rolls-Royce Ltd*

*Figure 11.2. Photograph of test-bed and preparation area used in conjunction with the development of the RB-211 engine: blades shown are made out of carbon fibre. Courtesy Rolls-Royce Ltd*

rain and sand erosion of the carbon-fibre blades. The difficulty has been overcome by the incorporation of a stainless-steel sheet leading edge which is moulded in during blade manufacture. At the time of writing, problems still exist with impact resistance to large objects such as birds, but it is expected that these difficulties will be overcome in the future. Figure 11.1 shows the RB-211 engine, which for the time being is fitted with titanium blades[23-26], while Figure 11.2 shows its test-bed area.

Apart from the rotating fan blades, other aero-engine parts suitable for the use of carbon fibre are the outer engine cases and retaining rings. If the development of metal-matrix composites proves successful, they will be used for the higher temperature blades and components, but progress to this stage will take a number of years to reach. Very substantial development is also being carried out by US aero-engine manufacturers to exploit the use of advanced composites based on both carbon and boron.

### 11.2.3   AEROSPACE

Tremendous strides have been made in aerospace during the last few years, culminating in the landing of US astronauts on the moon in the Apollo 11 mission. Rocketry, whether for the exploration of space or military uses, demands the utmost performance from every component and from each part of the structure. Success in the exploration of aerospace depends, possibly to a greater extent than in any other field, on the ability to minimise weight requirements. The answer is not to be found simply in the use of larger and more powerful motors, since these require more fuel which brings one back into the vicious circle, but depends on lighter materials of construction. Advanced composites have already started to play a major role in the game of weight saving, where every pound or fraction of a pound in weight saved may spell the difference between success or failure.

Composites, and indeed any new material of constructon, must be subjected to very thorough proving before use in an actual aerospace vehicle. Thus, lengthy ground testing and evaluation must be carried out first if an accurate assessment of properties, consistency, and reliability of the material is to be obtained. Nevertheless, the scope for composites including those containing carbon fibre is considerable, the main areas of application being:

| | |
|---|---|
| Re-entry vehicle structures | Rocket structures |
| Rocket motor cases | Satellite structures |
| Rocket nozzle reinforcement | Tanks and pressure vessels |

Much of the work being carried out in aerospace is of a highly confidential nature and hence it is difficult to give detailed information. The published work is largely confined to general description and indication of likely areas of application.

In the UK much work has been carried out by Imperial Metal Industries Ltd, Summerfield Research Station[27]. Thus, IMI, at the invitation of the Ministry of Technology, constructed a satellite frame out of aluminium honeycomb covered with a skin of CFRP. This experimental frame had much the same weight as an all-alloy one, but with the experience gained from this work it is believed that weight reductions of up to approximately 30% are feasible. IMI have also developed advanced techniques for the manufacture of tanks and bottles using filament winding in which glass fibre or carbon fibre can be used, for the most part in conjunction with epoxy resins. Figure 11.3 shows a photograph of the satellite structure. Figure 11.4 illustrates a pressure bottle made by filament winding, while Figure 11.5 is a photograph of a fuselage section also made by filament winding.

*Figure 11.3. A satellite structure made from CFRP sections. Courtesy Imperial Metal Industries Ltd*

*Figure 11.4. A filament-wound pressure bottle. Courtesy Imperial Metal Industries Ltd*

*Figure 11.5. A fuselage section made by filament winding and incorporating stiffening rings and stringers. Courtesy Imperial Metal Industries Ltd*

Also in the UK, Bristol Aerojet Ltd have investigated the potential of carbon-fibre composites for rocket motors and allied structures. Trigg[28] has indicated that, whereas low-modulus fibre already finds use for ablative applications, the high-modulus form is being developed for rocket motor cases using filament-winding techniques similar to those obtained by IMI. Other structural components are also being considered.

A wide range of activities connected with carbon-fibre composites has been undertaken in the USA by such organisations as NASA and the US Air Force. Many private companies are also involved. Their programmes of work range from the truly theoretical to the fabrication and testing of hardware. Kennedy[29] of the Boeing Co. has indicated his company's interest in high-modulus carbon-fibre composites for structural elements in rockets. In particular, the Boeing Co. have designed, fabricated, and tested the components for a missile-interstage shell structure which fits between the second and third rocket stages. Kennedy indicates that, from the property data on a cylinder, a hat-section stiffener, an attachment joint, and access-port reinforcement elements, it is likely that in the whole interstage structure a weight saving of about 27% can be achieved.

Although the aerospace sector will make a major contribution to the science of composites and facilitate their wider use, all the indications are that the quantities involved will be relatively low; certainly much lower than for aircraft. If, however, the price of carbon fibre becomes significantly reduced, there is scope for reasonable quantities of advanced composites for use in applications such as military rockets.

## 11.2.4    SUMMARY

Progress in the development of advanced composite materials in aircraft structures, aircraft engines, and aerospace fields has highlighted a number of common factors which are enumerated below:

1. Weight saving is of paramount importance.
2. Structural parts and components involve very advanced materials and technology.
3. The cost of materials used, whilst important, is much less critical than for most other technological areas. As a result, advanced new materials such as carbon-fibre composites are being evaluated for many applications.
4. Much exhaustive testing will be needed on any particular composite component before use in quantity can begin.
5. Existing carbon fibre properties are adequate, but good

consistency of mechanical properties and handling characteristics will be demanded.

The above factors have combined to make the aircraft/aerospace
market the most important at the present time, where virtually all
of the output of carbon fibre is being used. Full production programmes are still some way off, possibly two or even four years, but
the test data being established now will decide the extent to which
carbon-fibre reinforcement will be used in this industry. Today,
prices are high compared with conventional materials, but already
they are falling and will continue to do so as higher production
levels are achieved, commensurate with truly commercial usage in
aeroplanes. This, in turn, will help to provide the platform necessary
for applications outside aircraft and aerospace in the form of lower-
cost fibre and a wealth of composite materials technology.

## 11.3   OTHER APPLICATIONS

Preliminary work of an exploratory nature is being carried out in a
number of areas, namely:

| | |
|---|---|
| Automobiles | Marine engineering |
| Chemical engineering | Mechanical engineering |
| Electrical engineering | Sports equipment |

In each of these fields, the work is at a very early stage of development, but progress will be greatly assisted by the results attained
in the aircraft and aerospace sectors. In the case of sports equipment,
technical data and testing will be required but the final proof of
success will depend on the likes and dislikes of sportsmen.

### 11.3.1   AUTOMOBILES

At the present time, the case for using appreciable quantities of
carbon-fibre composites in normal production cars is simply not
valid. Car manufacture is highly competitive, and existing fully
established constructional materials are perfectly adequate. Even
the use of glass fibre for body construction is not a commercial
proposition unless the car is of a semi-special type having limited
production runs. Where the volume of manufacture is large, sheet-
steel bodies can be produced more economically, a factor which
outweighs other considerations. For carbon fibre to become established in the average family saloon, it must either show some outstanding advantage or be more cost effective.

Although the above remarks apply very strongly to the average production car, very high quality sports cars and racing cars present a rather different picture. In the case of a high-performance sports car, the cost of manufacture is high in the first place owing to advanced design, better than average workmanship, and small production runs. Hence the concept of increased performance due to the incorporation of a material like carbon fibre makes sense. In racing cars, the picture is again different; a racing car is built to win races and, if this can be helped by weight saving, increased structural strength, the possibility of a more advanced aerodynamic design, or better road-holding, then the use of carbon-fibre reinforced plastics is fully justified. Today, much thought is being given by the makers of racing cars to the possibilities of new materials including advanced fibre composites. In a paper presented by Tsai[30], details are given of the properties of boron-reinforced epoxy-resin composites for body applications.

Currently, carbon-fibre reinforced nose sections for the Lola GF type T70 are being moulded by Specialised Mouldings Ltd[31] from carbon fibre supplied by Courtaulds Ltd. On the Lola car body, 2 lb of carbon fibre saved 30 lb total weight, which represents a reduction of 20% in body weight.

Carbon-fibre reinforcement has also been used by Glass Fibre Engineering Ltd in the construction of the Ford GT40 body[32]. This was made essentially from glass fibre–resin, and a grid of British PAN-based fibre tows of 10000 filaments was incorporated, the fibre being supplied by Courtaulds and by Morganite Modmor. The weight of the carbon fibre amounted to less than 1 lb, but the resultant gain in strength allowed a body to be made having a thinner section, giving a weight saving of over 50 lb; moreover, the body was actually stronger than the original one and had greater freedom from vibration.

Thus, carbon fibre has already proved itself in a racing car application as the car in question won the race at Le Mans. Other possible areas of application are numerous, but much design and evaluation will need to be carried out before some potential uses become a reality. The *Automobile Engineer*[33] has reviewed the use of carbon fibres in vehicle design, giving possible applications such as bearing materials, connecting rods, push rods, propellor shafts, axle casings, and even spring hanger bearings. It is stated that, owing to the relatively high cost of carbon fibres, they will be used first in applications where a small quantity can be put to good advantage. Indeed, this is exactly the case with the Ford GT40. Tidbury and Tetlow[34], of the Cranfield Institute of Technology, gave data relevant to the use of carbon fibre for vehicle structures,

H

stressed the need for improved load predictions on automobile structures, and discussed methods of load calculations. They agree that in the foreseeable future carbon fibre is unlikely to be used widely for vehicle structures because it is expensive and, what is of considerable importance, because its composites do not yield under load as do steel components. Thus, a composite structure is not able to absorb impact in the event of a crash to the same extent as metals. Furthermore, joining carbon-fibre composites to other materials may introduce some weight penalty and lower the overall saving.

In the field of road transport, composites may find application, possibly in the form of GRP which may be combined with carbon fibre to upgrade its properties. Based on the principles of the GT40, it is not difficult to visualise a container or tanker in which GRP is the main component, suitably reinforced with carbon fibre, possibly in the form of a grid to impart increased stiffness and hence a reduction of the total structural weight so that the payload can be increased; any significant increase in payload can justify the extra cost, particularly if amortized over a number of years. Nevertheless, much work needs to be carried out in this field before one can expect to see wide use of CFRP[35].

## 11.3.2 CHEMICAL ENGINEERING

High-modulus composite materials are likely to find specific applications in the chemical engineering field. As in the case of the automobile industry, the areas selected will be those where a marked degree of improvement can be made. It is considered that carbon fibre will not find use at an early date in tanks and pressure vessels, for instance, since, in a static installation, weight saving does not give any benefit and conventional metals will be very difficult to compete with on price. A number of areas does exist, however, where the high specific stiffness of carbon fibre can be put to good advantage.

A possible area of application is for the shafts of stirrers and agitators which are widely used in the industry. Where a reaction vessel is very deep, an agitator will probably require a bottom bearing if whipping of the shaft is to be prevented at high rotational speeds. Submerged bearings are very costly, and the alternative is to provide a very heavy shaft which, in turn, needs larger top bearings and more power to drive. A similar situation exists with pump shafts installed in vessels, and both applications could make the best use of carbon-fibre composites; moreover, the weight of fibre required is relatively small, giving a high probability of cost effectiveness.

With regard to chemical inertness, both glass-fibre and carbon-fibre composites frequently show similar resistance to chemical attack, but usually the resin properties are the determining factor. In the case of strong alkali solutions, it has been shown that carbon has the advantage over glass, which is readily attacked; moreover, carbon is not affected by prolonged immersion in water. Today, however, relatively little information is available on the behaviour of carbon-fibre composites in contact with chemical reagents although information exists on resins, as indicated in Chapter 10.

Other areas where the high specific strength, lightness, and chemical inertness of carbon-fibre composites may be used to good advantage is in the manufacture of high-speed fans, centrifuges, etc. Again the weight of fibre and hence cost are likely to be low in comparison with the benefits derived. These and other possibilities are discussed in References 36–41.

## 11.3.3   ELECTRICAL ENGINEERING

In this field, a number of possibilities exists for utilising the outstanding mechanical properties of carbon fibre. In one of these, use is made of carbon fibre to reinforce a metal; thus, in electrical transmission lines, the fibre strengthens the copper or aluminium conductors and allows greater distances to be used between support towers. It is very doubtful whether normal transmission lines will ever justify carbon fibre incorporation, but, in cases where extra long spans are necessary as in river crossings, its use may prove advantageous and justify the extra cost. In the development of super-high-speed electric trains, the overhead conductors must be as free as possible from sagging and from deflection by the upward force of the pantograph, otherwise excessive wear and arcing will take place. In this case, there is a very real incentive to provide overhead lines of the maximum rigidity, and carbon fibre may be used to reinforce the copper conductor, the latter also acting as the matrix material. It is likely to be many years, however, before widespread use is made in these areas.

A different area is associated with high-speed rotating machinery such as alternators where the trend is to design larger and larger units to give improved operating efficiency. In the largest alternators, the mass of the copper bars in the rotor is such that very large centrifugal stresses exist at normal running speeds, and the stage has been reached where conventional materials of construction will not allow larger diameters. The problem is made worse by the close gap which must be maintained between rotor and stator, and any creep may prove catastrophic. Use of larger alternators is beset by the problem of shaft whip and does not offer a solution. Carbon fibre may find application for retaining bands on the rotors of such machines, but much experimental work will be necessary to establish this application. Nevertheless, carbon fibres are being evaluated at the Nelson Laboratories of English Electric–GEC[42] as a material to replace steel in the bell ends (or retaining rings) of alternator rotors. Carbon fibre supplied by both Courtaulds and Morganite Modmor has been used in experimental work in this field. It is believed that, whereas metals can just be used with alternator capacities up to 1300 MW output, designs for even greater output levels will need better materials; hence the interest in carbon-fibre composites.

In the area of electronics and sound reproduction, one possible application is for high-quality loudspeaker cones, where the combination of high stiffness and low weight are ideal for improved frequency response. The weight of fibre used would be small, making little difference to the cost of the speaker.

Work has recently been undertaken on the use of carbon fibre as a brush material in electric motors. Bulk carbon has been used almost exclusively for this operation for many years, but it is now believed that carbon fibre may prove better in certain cases, particularly in conjunction with electronic switching. However, the work is still at the early stages of development.

### 11.3.4   MECHANICAL ENGINEERING

Numerous areas of potential application exist for carbon-fibre composites, but cost effectiveness will be necessary as in the other fields. Aspects of mechanical engineering which are of particular interest at the present time are:

Bearings
Components and structures for fast surface transport such as railways
Gears
High-speed rotating machines
High-speed textile machinery
Machine tools
Medical applications
Structures such as masts, towers, cranes, etc.

In the area of transport, British Railways[43] have carried out work with carbon-fibre gear wheels which are able to run without lubrication; the fibre was supplied by Courtaulds. Oil seepage has been a notable problem with diesel-electric locomotives, and CFRP gear wheels have been shown to be capable of running without lubrication whilst still retaining good life. The carbon fibre was used to reinforce nylon in amounts up to about 15% by weight; as a result, the overall cost was reasonable. The success depends on the combination of reinforcement and lubricating qualities of the fibre which allows the nylon to work under heavier loadings (see p. 155). Further work with reinforced-plastic gears is being undertaken by the Production Engineering Research Association (PERA)[44], where a range of plastics reinforced with carbon, boron, glass, and asbestos is being evaluated. One aim is to produce machine-tool gear boxes which will be much less noisy and which will not require lubrication. In reinforced-plastic gear wheels, the amount of fibre needed is small, and therefore cost effectiveness appears to be attainable even today. In the area of high-speed rail transport, it is possible that carbon fibre in conjunction with glass fibre may find use on structural members and body work, but the main problem will be the cost of the carbon fibre. A further application may be in the making of tools sets for the pressing of plastics where the excellent dimensional stability of carbon fibre composites together with their resistance to wear, near-zero coefficient of expansion, and greatly reduced weight can be used to advantage.

High-speed weaving is a promising area for composite materials, and Bonas Brothers[45] are developing looms capable of working at twice normal speeds as a result of using aluminium reinforced with carbon-fibre composites in place of steel for heddle frames. In this work, the fibre was supplied by Morganite Modmor and Courtaulds to Fothergill & Harvey who made the frames. PERA[46] have produced a carbon-fibre bow for wire-stranding machines in conjunction with Trafalgar Engineering. It allows higher operating speeds and uses Courtauld's 'A' fibre and CIBA resins. Clearly, any saving in weight and inertia will enable higher speeds and profit-earning capacity to be obtained for both these applications; moreover, each application uses a small amount of an expensive material to effect a marked improvement in performance not attainable by other materials. The use of lightweight composites may be considered for other high-speed reciprocating machinery to reduce inertia and increase performance.

Towers and masts are obvious applications for composites where their low weight coupled with high strength and stiffness allows cost savings and greater safety factors or greater heights to be achieved. Equally, such applications as radar aerials are attractive in view of the need to maintain very close dimensional tolerances on large moving structures.

One interesting area combines mechanical engineering and medical applications. North American Rockwell Corp.[47] have found that certain carbon-fibre composites are chemically, biologically, and physically compatible with the fluids and tissues in the human body. Most metals, however, are not, either setting up local irritation or suffering from corrosion. Thus it is possible that bones, joints, etc., may be partly or wholly replaced by CFRP, but, before any widespread use can arise, considerable effort must be spent on trials.

### 11.3.5   MARINE ENGINEERING

The main areas are:

|                          |                    |
| ------------------------ | ------------------ |
| Hovercraft and hydrofoils | Sailing-boat masts |
| Hydrospace               | Ships' propellors  |

Hovercraft bear some resemblance to aircraft in that structural weight is an important factor if maximum payload is to be achieved. Hovercraft depend on air pressure from fans to maintain a small clearance from the water so that, in effect, mechanical energy is used to provide the lift as in an aeroplane. Additional advantage may be derived from fabrication of the hovercraft propellers from carbon-fibre composites to save weight and to reduce vibration by means of the higher damping factor of these materials relative to metals. As in so many of these applications, it is likely that either the carbon fibre will be used in conjunction with glass or alternatively the composite will be bonded to metal skins and structural members to provide stiffness in critical areas.

Sailing-boat masts are a natural application for carbon fibres since the main problem is one of stiffness relative to weight, and CFRP can provide an ideal solution. In order to derive the maximum benefit from a sailing boat, the sails must be attached to a stiff mast which does not bend and whip in strong winds, otherwise the angle of the sail alters and performance suffers. One method of using carbon fibre is to bond an outer skin on to a metal mast; in this way, considerable increase in stiffness is obtained at minimum fabrication and fibre cost. However, for the maximum performance an all-carbon mast will be superior but the extra cost may be difficult to justify. A report from the University of Southampton[48] deals with modern developments in materials for yacht construction and suggests areas where GRP and CFRP may be used. Of particular interest are the diagrams given in the report, some of which show how the end attachments are made with carbon-fibre composites.

Ships' propellers at first thought do not present a viable application for carbon fibre since metal has ample strength and is cheap.

However, the modern trend towards super-large tankers has meant that large propellers may weigh 30 tons or more which have to be supported by the aft bearing of the propeller shaft. The bearing and shaft have to be very large, and friction absorbs a considerable amount of power. At the present time, large tankers have two screws; if a single screw could be used, only one propeller shaft and one reduction gear set would be necessary, which would save cost, weight, and power. In metal, however, the weight would be too great for a single bearing, but not if the fabrication were carried out with carbon-fibre composites. Such an application is very unlikely within the next few years, but in the next decade large propellers may be given serious consideration.

Hydrospace implies the exploration of the sea, often at great depths. Whereas metals have given excellent service for the construction of hulls, etc., the quest for service at greater and greater depths has created considerable difficulties with metals owing to the problems of weight. The hull of a submarine, for example, is subjected to large compressive forces which are proportional to the depth, and, to withstand these forces, the hull thickness has to be increased. A point is reached where the law of diminishing returns rapidly sets in since the hull becomes so thick that all buoyancy is lost and the payload is reduced to zero. Moreover, the welding of metal of such thickness is costly and flaws are likely to occur.

As a result, interest is being shown in GRP for submarine hulls, the intention being to fabricate a whole hull by filament-winding techniques, the material being much lighter (for equal strength) than steel. Carbon fibre may be a candidate for such an application in view of its high mechanical strength and ability to withstand prolonged immersion in sea water without deterioration. To be successful, however, the fibre will have to be available at prices comparable to existing materials of construction and it is likely that military craft will act as a spur to progress.

## 11.3.6  SPORTS EQUIPMENT

The last two decades have seen a notable increase in the degree of 'professionalism' in the sports world, and in almost every field of sport there are financial incentives. It is logical to suppose, therefore, that the professional sportsman will demand the very best from his equipment. Already a number of applications have been established where CFRP may provide improved performance as a result of strength or stiffness coupled with low weight. Indeed, it is fair to state that sports equipment may prove to be one of the first and also largest users of carbon-fibre composites. Examples are:

Bows and arrows
Camping gear
Canoes
Fishing rods
Golf clubs
Oars
Racing car bodies (see Section 11.3.1)
Skis
Tennis rackets, squash rackets, and cricket bats
Yacht masts (see Section 11.3.5)

In this field the object is to win, and, if better equipment makes this easier, extra cost can be justified. For example, a golf shaft which is stiffer without an increase in weight will permit a greater drive, which is of supreme importance to the player intent on winning the game. Already, golf clubs, tennis rackets, and the like have been constructed out of carbon-fibre composites. The next stage must be a critical evaluation in the hands of experienced users for it is only they who can decide if the club or racket has the right 'feel' in their hand. The fullest laboratory testing will be meaningless unless the user approves and likes the way in which the equipment performs.

Even at the current high prices, there can be little doubt that carbon fibre may be well and truly cost effective in these areas. This fact, coupled with the size of the sports-goods market, may well prove to be a potential outlet in the short term for carbon fibre comparable or even greater in size than the aircraft industry.

An indication of the interest being shown in sports equipment is given by Paton[49], who looks at designs where CFRP is used. Gunn and Moore Ltd[50] have developed a new cricket bat in which the conventional steel spring in the handle was replaced by one of CFRP; this produced a weight saving in the handle and an increase in the weight of the blade resulting in improved balance, an increased area of accurate strike, and a more powerful drive. The fibre was supplied by Courtaulds, and Nottingham County Cricket Club is to be amongst the first teams to switch to the new bats.

Camping gear does not normally have a competitive edge and for average use it is doubtful if existing materials will ever be ousted by composites. For military applications, however, the position could be entirely different, and the saving of weight or space or the ability to generate greater efficiency may prompt their use. This is particularly true for equipment which has to be air freighted or is used for mountaineering.

## 11.4   DESIGNING IN CARBON-FIBRE REINFORCED PLASTICS

### 11.4.1   GENERAL OBSERVATIONS

In the preceding chapters, the properties of carbon fibres were discussed in relation to other bulk and fibre materials. Composite bodies of carbon fibre and resins (CFRP) were considered, and stress was laid on the characteristics which have led to current applications. It must be recognised, however, that detailed knowledge of composite properties is still limited, in contrast to the extensive data on metals and unreinforced plastics.

The engineering designer faced with the design of a new machine has options on many materials of construction. His is the responsibility of selecting materials so that the end-product meets the requirements at the lowest manufacturing cost. In many instances, CFRP will be able to improve the performance or save cost or both, but, before embarking on a major development programme, the designer must know in broad terms if CFRP is suitable or not. He therefore needs guidance on the likely merits of CFRP versus other materials.

The cost of a material of construction must be a critical factor in deciding its use. Cost, however, is not the only factor, and failure to recognise this may be a short-sighted policy, as illustrated by an example: a fan manufacturer wishes to make a fan having a certain performance level and an upper limit is set for manufacturing costs. One method is to use the cheapest viable material for, say, the blades; a second method is to use more expensive blades fabricated from a lighter and stiffer material such as CFRP, thereby saving weight and permitting a smaller shaft and bearings. The overall result may be a nett saving in manufacuring cost and a more efficient fan. Furthermore, if the CFRP blades prove to be more free from vibration, benefit will be felt in the whole fan and ducting system as a result of lower noise levels, and the bearings may have a longer life. These may be vital selling points and increase sales and profits.

Thus, initial material cost is not necessarily the all-important factor. It is the overall cost effectiveness which counts as measured in terms of: (a) total manufacturing cost; (b) overall performance; (c) reliability, and (d) ability to perform better than competition.

At the present time, carbon fibre costing, say, £50/lb cannot stand comparison with steels or other conventional metals and for this reason the designer may think that he should not get involved. However, special metal alloys such as titanium costing several pounds per pound weight are finding more widespread use in appropriate applications, where they can show overall cost effectiveness. Since carbon fibre has four times the specific stiffness and strength of steel, its effective price is reduced proportionally; moreover, the cost of carbon fibre has already been reduced considerably as the volume of production increases and usage becomes more closely associated with manufacture rather than development or prototype testing. If one makes allowances for the time-scale involved in the introduction of new materials into any application, it is certain that fibre prices will be further reduced by the time large production runs are contemplated. Now is the time to start work on applications deemed suitable for taking full advantage of the outstanding properties of CFRP.

## 11.4.2   DESIGN CONSIDERATIONS

With conventional materials much expertise has already been established, not only in their properties but in methods of fabrication, testing, reliability, availability, and cost. With composites, or any new materials, such a baseline does not exist.

In considering the suitability of composites such as CFRP, the designer has to start right back at the beginning. He has to ask himself if CFRP is suitable at all, and how to select the most likely area of application. Bedwell[51] has outlined the requirements which must be fulfilled if CFRP is to be found suitable for any particular application. He states that it should satisfy at least one of the following requirements:

1. It should possess a mechanical robustness greater than that presented by unreinforced plastics and must have a higher resistance to chemical attack than that of GRP, the latter tending to delaminate in alkalis and to some extent in water.
2. It should have self-lubricating and improved wear characteristics relative to unfilled thermoplastics and be able to withstand higher loads or rubbing speeds than bulk carbon.
3. It should possess the maximum specific modulus coupled with high specific strength.

In addition, CFRP must be cost effective in the ultimate analysis for any application. Bedwell has prepared a chart showing the essential questions and answers involved in selecting CFRP for a particular application. This is given in Figure 11.6.

Having selected a likely application area, a host of supplementary questions follows, namely: What are the kinds of load, static or dynamic? How is the composite to be attached to other parts of the

structure? In what directions do the loads act upon the CFRP structure? What about fatigue? Is corrosion a problem? Is stiffness much
more important than strength? What about cost and ease of fabrication? Is toughness vital?

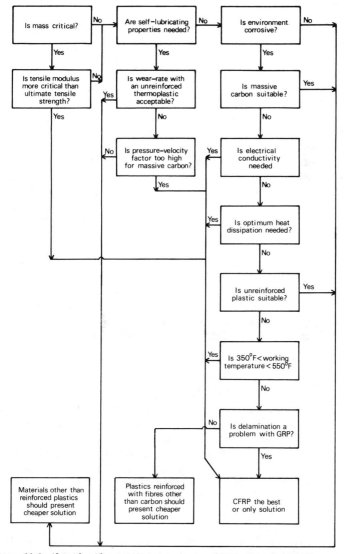

*Figure 11.6. Algorithm showing steps in selecting CFRP. From Bedwell[51], courtesy
Morganite Research & Development Ltd*

These and many other questions will have to be answered, but it is only by working through the problem in a systematic way that the answers can be assembled. A likely solution having been reached, the next stage is to make the component and evaluate its performance.

It is likely that a certain amount of trial-and-error work will have to be carried out before the final solution is found. One may find that the structural part is different in shape from one constructed out of conventional materials, and this should not be allowed to inhibit design progress in composite materials.

Final testing may take place by merely carrying out direct substitution, but, for aircraft structures or others whose safety is critical, this may not be possible. In such cases, testing will have to be done first in non-critical areas or may be preceded by ground testing, as appropriate.

Build-up of data both from the engineering side and from the commercial cost effective viewpoint may take years, but ultimately a sufficient fund of knowledge will become available to allow an ever-widening range of applications to be tackled, and this, in turn, will be coupled with the availability of fibre and composites at lower prices.

It is almost certain that the opportunities for CFRP will begin with applications where materials already in use have reached their limits of performance, as, for example, in the aircraft industry. A baseline will thus be established from which additional applications will follow, depending on how well the properties of CFRP can be utilised and recognising that material cost is only one of the factors which determine the overall cost and performance of the end-product.

### 11.4.3   USE OF CARBON FIBRE IN ASSOCIATION WITH OTHER MATERIALS OF CONSTRUCTION

It has been shown that the most important advantages of carbon fibre composites over metals or GRP result from their high values for specific strength and specific stiffness coupled with very good fatigue strength. At the present time, the cost of carbon fibre is high compared with metals or GRP, but in the fabrication of a component it is possible to replace a proportion of the lower-priced material with carbon fibre to obtain improved properties. A particular advantage of this approach is that the carbon fibre may be placed where it is most effective.

*Table 11.2* indicates the relative mechanical properties of a glass-fibre composite in the form of a mat or rovings, compared with the high-strength form of PAN-based carbon fibre with aligned

**Table 11.2** RELATIVE MECHANICAL PROPERTIES OF COMPOSITES MADE WITH E-GLASS AND CARBON FIBRE

| Materials | Tensile modulus ($10^6$ lbf/in$^2$) | Specific gravity | Specific tensile modulus ($10^6$ lbf/in$^2$) |
|---|---|---|---|
| E-glass fibre composite (glass 50% by volume | 2·0 | 1·66 | 1·2 |
| Carbon-fibre tows with parallel alignment (high-strength fibre 50% by volume) | 18·0 | 1·53 | 11·8 |

tows. It will be seen that the specific stiffness of the carbon fibre product is 11·8/1·2 = 9·7 times that of the glass-fibre product. It may be supposed therefore that 1 lb of carbon-fibre composite can always replace 9·7 lb of glass-fibre composite in a component such as a beam. This is only true if the thickness of the carbon-fibre composite cladding is a small fraction of the total depth of the beam. Where the relative thickness of the carbon-fibre cladding is large, some of the material is used ineffectively, by virtue of the 'depth cube' law for beam strength, and the overall advantage is reduced.

The amount of carbon fibre to be used in any particular application depends on the price that may have to be paid to save weight. *Table 11.3* gives details of the weight and cost of a series of GRP plates clad on both sides with high-strength carbon-fibre reinforced sheet. The latter is assumed to have a modulus of 18 × $10^6$ lbf/in$^2$ and a fibre content of 60% by volume. All the sheets in the series have the same stiffness. Two price levels for the carbon fibre are assumed for comparative purposes.

It will be seen from the table that a small addition of carbon fibre gives the greatest benefit. Thus, a 5% addition allows a 22% weight saving to be made; in contrast, 50% addition gives only a 52% weight saving. The reason for this is associated with the depth cube law as applied to the plate. Hence it is possible to estimate the amount of weight saving in a component containing varying amounts of carbon fibre. These data may also be used to ascertain the overall cost of the component and indicate the optimum proportion of carbon fibre to be used.

The table refers to the problem of replacing GRP with carbon-fibre composite to give components of the same stiffness but of less weight. The same argument applies to metals such as aluminium where the addition of 10% of carbon-fibre cladding will allow 1 kg of weight to be saved at an extra cost of either £36 or £12, based on the two fibre prices given in the table.

In the case of aluminium and other ductile metals, it is always

**Table 11.3** EFFECT OF VARYING PROPORTIONS OF CARBON-FIBRE CLADDING TO GRP IN TERMS OF WEIGHT SAVED AND COST

| Fraction of thickness of plate formed from carbon-fibre composite sheet | Thickness of plate (cm) | Total cost of materials for plate (£/m²) | | Weight of plate (kg/m²) | Savings of weight made possible by the use of carbon fibre (%) | Cost of saving 1 kg of weight in the plate (£) | |
|---|---|---|---|---|---|---|---|
| | | Carbon fibre at £100/kg | Carbon fibre at £35/kg | | | Carbon fibre at £100/kg | Carbon fibre at £35/kg |
| 0 | 1·00 | 11 | 11 | 16·6 | 0 | 0 | 0 |
| 0·05 | 0·78 | 49 | 22 | 12·9 | 22 | 10 | 3 |
| 0·1 | 0·68 | 78 | 30 | 11·2 | 32 | 12 | 3·5 |
| 0·2 | 0·59 | 129 | 46 | 9·6 | 42 | 17 | 5 |
| 0·5 | 0·50 | 265 | 90 | 8·0 | 52 | 30 | 9 |
| 0·7 | 0·48 | 359 | 119 | 7·5 | 55 | 38 | 12 |

Notes:
1. The carbon-fibre cladding is in the form of sheets containing 60% by volume of fibre.
2. The glass-strand mat consists of randomly oriented fibres of 35% by volume and costing £0·8/kg.
3. In each case the resin matrix is an epoxy.
4. All plates are of constant stiffness and are considered to be bent in one direction only.

most profitable to use carbon-fibre cladding to increase the strength
of the components, quite apart from the stiffness, since the difference
in their respective specific strengths is usually large. This favours
the use of carbon fibre in conjunction with ductile metals.

The information presented in the above section is intended only
to serve as a guide to show the possibilities of combining carbon
fibre with conventional constructional materials in order to enhance
their properties. In practice, each type of component must be
treated separately and, if necessary, trials will have to be carried out
in order to attain optimum properties. If carbon fibre becomes
available at around £7/kg, it will be possible to reduce the weight
of aluminium structures without increased cost.

### REFERENCES

1 ANON., 'World Plastics – Financial Times Survey', *Financial Times*, No. 24873, 11 (16.6.69)
2 MORGANITE RESEARCH & DEVELOPMENT LTD and WHITTAKER CORP., NARMCO MATERIALS DIVISION, 'Modmor Carbon Fibre', *New Scient.*, **43** No. 660, 240 (1969)
3 MCELHINNEY, D. M., KITCHENSIDE, A. W., and ROWLAND, K. A., 'Carbon Fibres Take to the Air', *New Scient.*, **43** No. 661, 282 (1969)
4 HIERONYMUS, W. S., 'Carbon Composite Program Gains', *Aviat. Week*, **91** No. 7, 52 (1969)
5 LTV AEROSPACE CORP., 'Graphite Fibre Wing Tip', *Iron Age*, **204** No. 6, 27 (1969)
6 ANON., 'Graphite and Boron Composites', *Am. Mach., N.Y.*, **113** No. 16, 39 (1969)
7 NORTH AMERICAN ROCKWELL CORP., NORTH AMERICAN AVIATION DIVISION, 'Plastics-Reinforced Honeycombs Help Form Bonds in One Step', *Product Engng*, **40** No. 17, 62 (1969)
8 UNION CARBIDE CORPORATION, CARBON PRODUCTS DIVISION, *Thornel Graphite Yarn*, Technical Information Bulletins 465-202-GG, 465-203-GG, 465-204-AH, 465-205-EH, 465-206-BI (1966–68)
9 US AIR FORCE, 'Advanced Composites Move Closer Reality', *Mater. Engng*, **70** No. 4, 19 (1969)
10 MINISTRY OF TECHNOLOGY and FOTHERGILL & HARVEY LTD, 'Carbon Fibre Keeps Jets in Good Trim', *Engineer, Lond.*, **229** No. 5838, 7 (1969)
11 WESTLAND HELICOPTERS LTD, 'Carbon Fibre in Aircraft', *The Times*, No. 57735, 25 (5.12.69)
12 JEFFS, E., 'Westland's Way with Carbon Fibres', *Engineering, Lond.*, **208** No. 5407, 614 (1969)
13 SCHWARTZ, N. B., 'Composites: They're Flying High', *Iron Age*, **204** No. 14, 83 (1969)
14 GRESHAM, H. E., 'Materials Aspects of Advanced Jet Engines', *Metals & Mater.*, **3** No. 11, 433 (1969)
15 WHITNEY, I., ROWLAND, M. R., and JONES, S. G., Second Prize, *New Scientist Award* (1968)
16 ROLLS-ROYCE LTD, Germ. Pat. 1 940 306 (26.3.70)
17 MORTIMER, J., 'Now – Mass Production of Carbon-Fibre Parts', *Engineer, Lond.*, **230** No. 5947, 7 (1970)
18 MORTIMER, J., 'Automatic Presses Speed Output of Fibre Blades', *Engineer, Lond.*, **230** No. 5951, 58 (1970)

19 GRESHAM, H. E., 'The Development of Fibre-Reinforced Composites for Gas Turbines', *Prod. Engr,* **48**  No. 9, 393 (1969)

20 ROLLS-ROYCE LTD, Fr. Pat. 1 575 682 (25.7.69)

21 ROLLS-ROYCE LTD, Fr. Pat. 1 578 956 (22.8.69)

22 ROLLS-ROYCE LTD, Fr. Pat. 1 579 649 (29.8.69)

23 ROLLS-ROYCE LTD, Brit. Pat. 1 170 592 (12.11.69)

24 ROLLS-ROYCE LTD, Brit. Pat. 1 170 593 (12.11.69)

25 ROLLS-ROYCE LTD, Brit. Pat. 1 186 486 (2.4.70)

26 ROLLS-ROYCE LTD, 'Titanium is Away by a Short Head', *Engineer, Lond.,* **230** No. 5955. 7 (1970)

27 IMPERIAL METAL INDUSTRIES LTD, SUMMERFIELD RESEARCH STATION, 'Satellite Frame from Carbon Fibres', *New Scient.,* **43** No. 659, 191 (1969)

28 TRIGG, T. A., 'Carbon Fibre Composites in Rocket Motor Systems', *J. Br. Interplanet. Soc.,* **22**, 337 (1969)

29 KENNEDY, P. B., 'Composite Structural Elements', *National SAMPE Tech. Conf.: Vol. 1, Aircraft Structures and Materials Application,* Western Periodicals, North Hollywood, 307 (1969)

30 TSAI, S. W., 'Plastics for Vehicle Bodies: A Progress Report and Survey of Recent Literature', *Automot. Des. Engng,* **8**, 44 (1969)

31 SPECIALISED MOULDINGS LTD, 'Specialised Mouldings to Build Wind Tunnel for Racing Car Body Development', *Reinf. Plast.,* **14**  No. 6, 155 (1970)

32 ANON., 'Prepregs and the Motor Industry', *Design & Compon. Engng,* No. 2, 20 (1970)

33 ANON., 'Carbon Fibres in Vehicle Design', *Auto. Engr,* **59** No. 13, 473 (1969)

34 TIDBURY, G. H., and TETLOW, R., 'Carbon Fibre for Vehicle Structures: Data Sheet 92', *Automot. Des. Engng,* **9**, 62 (1970)

35 SUMMER, J., 'Has Carbon Fibre a Place in Car Manufacture?', *Engineer, Lond.,* **230** No. 5953, 43 (1970)

36 ANON., 'Carbon Fibres: Aeronautical Spin-offs for CPI [Chemical Process Industry] Plants?' *Chem. Engng, Albany,* **76** No. 13, 36 (1969)

37 BARNABY, C. F., 'The Gas Centrifuge', *Sci. J.,* **5A** No. 2, 54 (1969)

38 HANSON, M. P., 'Tensile and Cyclic Fatigue Properties of Graphite Filament-Wound Pressure Vessels at Ambient and Cryogenic Temperatures' *Materials and Processes for the 1970's,* Society of Aerospace Materials and Process Engineers, North Hollywood, 249 (1969)

39 MORTIMER, J., 'Opportunity Knocks — but not for Long', *Engineer, Lond.,* **229** No. 5938, 5 (1969)

40 MINNESOTA MINING AND MANUFACTURING COMPANY, REINFORCED PLASTICS DIVISION, advertisement for advanced composite materials, *Mater. Engng,* **70** No. 4, 55 (1969)

41 FISHLOCK, D., 'Carbon Fibre for Centrifugal Stress', *Financial Times,* No. 25013, 11 (27.11.69)

42 PETERS, D., 'Carbon Fibres Can Spin through Generator Barrier', *Engineer, Lond.,* **230** No. 5957, 40 (1970)

43 BRITISH RAILWAYS BOARD, 'Carbon Fibre Gears Whirl Away the High-Cost Myth', *Engineer, Lond.,* **229** No. 5931, 35 (1969)

44 MORTIMER, J., 'PERA Goes into Top Gear with Fibre Testing', *Engineer, Lond.,* No. 5936, 10 (1969)

45 BONAS BROTHERS WEAVERMATIC LOOMS (ENGLAND) LTD, 'Carbon Fibres Speed Weaving', *Engineer, Lond.,* **229** No. 5938, 38 (1969)

46 PRODUCTION ENGINEERING RESEARCH ASSOCIATION OF GREAT BRITAIN and TRAFALGAR ENGINEERING LTD, 'Carbon Fibre Bow for Wire Stranding Machines', *Engineer, Lond,.* **230** No. 5962, 11 (1970)

47 NORTH AMERICAN ROCKWELL CORP., 'Carbon May Be the Next "New" Material for Surgical Implants', *Product Engng,* **40** No. 18, 62 (1969)

48 UNIVERSITY OF SOUTHAMPTON, DEPARTMENT OF AERONAUTICS AND ASTRONAUTICS, *Modern Developments in Materials Applicable to Yacht Construction,* SUYR Report 26 (1969)

49 PATON, W., 'Carbon Fibre in Sports Equipment', *Composites,* **1** No. 4, 221 (1970)

50 GUNN AND MOORE LTD, 'Carbon Fibre in Cricket Bats', *Composites,* **1** No. 4, 200 (1970)

51 BEDWELL, M., *Materials Towards the 70's,* Morganite Research & Development Ltd, London (1969)

# 12

# Future growth of the market for carbon-fibre composites

## 12.1 FUTURE GROWTH

Composite technology involving the use of high-modulus and high-strength carbon fibre is, as has been shown, three or perhaps four years old. It is only in the last two years that fibre of this type has been available in sufficient quantity to allow even large-scale research and development work to get underway. The field is therefore very much in its infancy. almost all the current output being used for evaluation programmes in the aircraft and aerospace industries, as stated in the previous chapter.

Although output levels are increasing and prices falling, the cost of carbon fibre is still much greater, even by orders of magnitude, than that of conventional materials. Hence application is today restricted to the higher technological areas able to derive the full benefit from this outstanding material. As usage and production levels increase further, there is no doubt that the manufacturing cost will continue to fall considerably and make attractive other areas of application, which, in turn, will lead to further price reductions and greater areas of usage. One starts at the top of the applications pyramid and works down to lower levels, each of these corresponding to increased usage. As one descends such a pyramid, each successive application layer will be funded by experience gained at the higher levels.

In looking at the future growth of carbon fibre and its associated matrix systems in whatever form they may take, it is useful to consider the growth of the reinforced plastics area as a whole. The reinforced plastics industry started roughly 25 years ago and has enjoyed a growth rate which few other industries can match. There is good evidence that in the USA in 1969 the total consumption of reinforced plastics amounted to almost $1000 \times 10^6$ lb. *Table 12.1* is taken from *Chemical and Engineering News*[1] and shows the growth of sales in this field from 1957 to 1970 in the various industrial sectors.

**Table 12.1** U CONSUMPTION* OF REINFORCED PLASTICS. From Parker[1], courtesy *Chemical and Engineering News*

| Market | 1957 | 1960 | 1963 | 1966 | 1969† | 1970† |
|---|---|---|---|---|---|---|
| Aircraft and aerospace | 25 | 30 | 35 | 46 | 42 | 53 |
| Appliances | 5 | 10 | 10 | 17 | 34 | 46 |
| Construction | 25 | 44 | 62 | 87 | 124 | 150 |
| Consumer products | 25 | 24 | 24 | 34 | 63 | 74 |
| Corrosion-resistant materials | 4 | 12 | 15 | 31 | 106 | 142 |
| Electrical components | 5 | 12 | 15 | 22 | 82 | 98 |
| Industrial equipment | 5 | 10 | 13 | 20‡ | | |
| Marine and accessories | 25 | 56 | 60 | 100 | 270 | 290 |
| Transportation | 34 | 45 | 54 | 100 | 220 | 318 |
| Miscellaneous | 15 | 12 | 10 | 25 | 53 | 63 |
| Totals | 168 | 255 | 298 | 482 | 994 | 1234 |

* Figures in the table are in millions of pounds and include the weight of reinforcement material.
† Estimated by *Chemical and Engineering News*.
‡ In 1966, industrial equipment was combined with appliances and electrical components.

It is likely that further growth rates will approach 25% per annum in many sectors of the reinforced plastics field; this has to be compared with 6–8% for the chemical industry in general and 4–6% for industry as a whole in the USA. Similar figures for the growth rates in reinforced plastics are likely to be generated in Europe in the 1970s, but allowance must be made for the two or conceivably five years delay in market fulfilment compared with the USA. Nevertheless, these figures are very impressive and show the enormous potential for reinforced plastics as a whole.

It has been shown that carbon-fibre composite materials possess exceptional specific strength and stiffness properties and currently fulfil the needs of a specialised part of the reinforced plastics market. Moreover, this situation will continue, certainly in the foreseeable future. It seems logical to believe, therefore, that carbon-fibre reinforcement will represent but a small fraction of the total market indicated above. Equally, it is almost certain that CFRP will always be more expensive than the vast majority of reinforcement materials used with today's multiplicity of plastics matrices which, in turn, will limit the area of application.

In taking a view of the potential for future growth during the 1970s, it is assumed that the total US market for reinforced plastics in 1970 is $1000 \times 10^6$ lb based on Table 12.1. If the growth rate is maintained in the 1970s, then the demand is likely to double by 1975 to $2000 \times 10^6$ lb and double again by 1980 to around $4000 \times 10^6$ lb. Current evidence of growth rates portrays such an expansion in the USA. A similar growth pattern will probably occur in Europe.

The use of carbon-fibre composites in the USA lies in the range 10 000–15 000 lb for 1970, corresponding to approximately $0.001\%$ of the total market for reinforced plastics materials. As applications for carbon fibre increase further, it is probable that by 1975 the percentage used will have grown significantly and will lie in the region of $0.01\%$ of the expanded total market. This implies a feasible consumption of 200 000 lb of CFRP. Similarly, by 1980 not only will the whole market have further expanded, but CFRP is likely to find still more usage, possibly extending up to $0.1\%$ which will give a consumption in 1980 of around $4 \times 10^6$ lb. These figures are extremely tentative as no-one knows how rapidly the advanced technology surrounding carbon fibres is likely to be generated and to what extent success is achieved in finding sizable applications. Very recently, a recession in the US market has occurred, particularly in aerospace, which is likely to delay this growth rate by at least two years.

Although $0.1\%$ share of the US market by 1980 represents but a very small fraction, it nevertheless indicates a tremendous growth rate over the decade and fully justifies the widespread belief that carbon fibres will play a leading role in the late 1970s and particularly in the 1980s in many forms of advanced material construction.

## 12.2   CONCLUSIONS

Metals have for many years been regarded as the only materials of construction where maximum mechanical properties are required. However, their capabilities have reached the stage where further marked increase in performance is not likely to be achieved. Since the demands for technological progress, particularly in aircraft and aerospace, are unlikely to be fulfilled by metals alone, a search for alternative types of constructional material has been under way for some time. The concept of using composites containing fibres of exceptional strength or stiffness is not new, but it is only in the last five years or so that suitable materials have become available, and of these carbon fibres are the most promising.

Work on the production of carbon fibres having a highly ordered crystallite structure, thereby allowing extremely high modulus values to be obtained, started in the mid 1960s. The first part of this work was centred on rayon-based or cellulose-based materials, the high modulus being attained by application of tension during the final graphitisation stages. Following this work, Watt and his colleagues at the RAE and also Rolls-Royce Ltd came up with a new approach which involved the use of polyacrylonitrile as the starting material. This gave several advantages over rayon: first, in that stretching was made to take place during the initial heat-treatment stage during which the fibre was also oxidised; secondly, a higher yield of carbon fibre relevant to the weight of precursor was obtained; and thirdly, the overall processing conditions were simplified to the benefit of commercial manufacture.

Of particular significance is the tremendous progress which has been made during the period 1965–1970. If this continues, one can confidently expect to see the manufacture of carbon-fibre based composite materials on a true commercial scale within the next few years.

The availability today of carbon fibre having consistent properties has allowed resin-based composites to be made with exceptional strength and stiffness. However, composites have a different structure and handling characteristics from those of metals and experience to date has indicated very strongly that, in order to make the most effective use of these materials, much design and applications work needs to be carried out. Already there is widespread activity in this area, particularly in aircraft companies, and it is confidently expected that carbon-fibre composites and associated technology will become established in the 1970s and begin to take their place alongside the vast range of metals, plastics, etc., which the engineer already has at his disposal. The vital necessity of thorough applications work and the preparation of reliable design data cannot be overstressed. Carbon fibre of today is satisfactory with regard to mechanical properties, but the materials can only be used to best advantage in the 1970s if the appropriate engineering data are generated.

## 12.3   POSTSCRIPT

Since the preparation of the manuscript for this monograph, an international conference on 'Carbon Fibres, Their Composites and Applications' was held in London in February 1971 under the auspices of The Plastics Institute. This conference provided a forum for the presentation and discussion of a very wide range of topics relating to the most recent studies on carbon fibres and their composites. In addition, it highlighted the widespread and growing interest in this field. It is appropriate, therefore, to make comment on the salient features of the conference.

The conference showed beyond doubt that widespread activity exists not only in the UK but also on the Continent and in the USA, Japan, and Russia. Out of a total of some 50 papers presented, about a dozen were concerned with the conversion of polymer yarns and filaments into carbon fibres. Considerable emphasis has been placed on gaining further knowledge of the chemistry of acrylonitrile-based precursors and the changes which take place when the material is decomposed to give a carbon-fibre end-product. This type of precursor still appears to show the greatest promise for the production of consistent high-quality carbon fibre for reinforcement. As a result of the work currently in progress on polymer decomposition mechanisms, structure, the role of defects, and the surface characteristics of carbon fibre, it is likely that significant advances will soon be forthcoming, in terms of both improved mechanical properties of fibre and greater consistency in their behaviour in composite structures.

A number of papers was presented dealing with the fabrication of composites, and it is noteworthy that much interest is currently being shown in metal matrix systems, either for ultimate use as high-temperature composites or for special applications, notably in electrical engineering. Nevertheless, resin matrices still predominate and progress is being maintained in establishing better understanding of their wetting and bonding characteristics with fibre. The conference indicated that much work is being undertaken on all forms of composite testing, particularly under specific environmental conditions, for example, deep-sea submergence. A range of composite data is being established for both thermosetting and thermoplastic resins, and of particular interest is the considerable work on carbon fibre–carbon matrix composites, which already appear most promising, particularly at very high temperatures. It should be noted that, from some of the data reported on composites, areas do exist where certain of their properties have failed to meet expectations based on theoretical data. This kind of experience is to be expected since it forms part of the 'learning curve' of composite fabrication and behaviour.

The final part of the conference was particularly relevant to the future of carbon-fibre technology. It dealt with the problems encountered in the design of CFRP components for engineering applications. Due reference was made to the need for the designer to 'think composites' in place of metals, and of the need for cost-effective solutions.

Above all, the conference highlighted the considerable and varied scientific work currently in progress in this field. It has shown areas of advancement and of difficulties experienced with these new materials. Nevertheless, in spite of the volume of testing and evaluation currently being undertaken, the use of CFRP in practical applications has been somewhat slower to materialise than envisaged two years ago. Part of this can be attributed to the relatively high cost of carbon fibres, but in the writer's opinion the main factor lies in the need to make more components and try them out under actual conditions rather than wait until full technical data have been established on test pieces. Some components will undoubtedly fail, but others will succeed, paving the way for wider use.

REFERENCES

1 PARKER, G. R., 'Reinforced Plastics 1970–75', *Chem. Engng News*, **28** No. 4, 57 (1970)

# Index